General Biology

Laboratory Manual

Fourth Edition

Charles A. Wade
C.S. Mott Community College

Wild mushroom image on cover courtesy of Faith Walker © Kendall Hunt Publishing Company.
All other cover photos © Shutterstock Inc.

www.kendallhunt.com
Send all inquiries to:
4050 Westmark Drive
Dubuque, IA 52004-1840

Copyright © 2001, 2005, 2015, 2019 by Charles A. Wade

ISBN 978-1-5249-8282-9

Kendall Hunt Publishing Company has the exclusive rights to reproduce this work,
to prepare derivative works from this work, to publicly distribute this work,
to publicly perform this work and to publicly display this work.

All rights reserved. No part of this publication may be reproduced,
stored in a retrieval system, or transmitted, in any form or by any
means, electronic, mechanical, photocopying, recording, or otherwise,
without the prior written permission of the copyright owner.

Printed in the United States of America

This book is dedicated in loving memory to Paul F. LeClaire (1938-2015).

Table of Contents

Acknowledgments

I would like to thank Margaret Clark and Melissa Wade for their critical reading and suggestions of this manuscript.

Introduction to the Laboratory

The laboratory manual is intended for non-science majors studying general biology. The biological terminology is kept at a minimum in an attempt to limit the technical jargon. The purpose of this laboratory manual is to expose the students to as many organisms and biological concepts as possible.

In some instances the laboratory exercises are very detailed while in other instances the laboratory exercises are very vague. This is because some of the organisms that will be covered are very familiar and others will be new to the students.

For most of the laboratory exercises the procedure will be as follows. There will be an introduction to the subject and materials by the instructor. The students will study the materials available in the laboratory. The manual will provide a guide and the specific objectives which specify what the student is expected to learn. Students may work in pairs or in larger groups as indicated by the instructor. A division of labor will frequently enable the students to accomplish more.

At the beginning of each and every laboratory period there will be a quiz. These quizzes will cover the material that was in the previous weeks laboratory exercise. These quizzes will be based on recognition rather than total recall. Function of structures will also be a large component of these quizzes. So, don't just memorize the material. Try to understand the material.

Mott Community College
BIOLOGICAL SCIENCES
Student Laboratory Rules and Procedures

General Guidelines

1. **Eating and drinking are prohibited in all biology laboratories.** Do not put on any lipstick or chapstick in the lab. Do not put pens or pencils that have been on the lab bench in your mouth.

2. Plan your lab work ahead, noting any special instructions and possible danger. Work defensively and cooperatively to assist fellow students.

3. Do not remove equipment, media, chemicals, models, microscope slides or glassware from the lab. The performance of unauthorized experiments is strictly forbidden.

4. Models and materials taken out of cabinets, tables or carts must have all parts intact when returned to the appropriate location.

5. Use only wood pointers or rulers to point out parts on bone, microscopes and models - never ballpoint or ink pen or pencil.

6. Read chemical labels very carefully. Take only what you actually need and never return unused reagents to the reagent bottle. Always screw down covers on all chemicals to avoid chemical spills.

7. All chemical waste and dissection material must be disposed of in the appropriate labeled waste container. Your instructor will point out the necessary containers for each lab.

8. Never look directly into ultraviolet light sources nor point them at classmates.

9. Following experiments, workbenches must be cleaned with bleach wipes.

10. Close all cupboards and drawers after use; chairs and stools should be pushed under the bench. Rinse experimental materials and return to the workbench for the next class period. Leave the lab the way you found it.

11. Wash hands thoroughly before leaving the lab.

Procedures

1. Laboratory experiments have been carefully designed for your educational benefit but also your safety. Follow written and verbal instructions **exactly** as they are given. Do not rush ahead during dissections or experiments; this is for your benefit.

Computer and Microscope Use

1. Use of college computers signifies acceptance of the Acceptable Use Policy. Computers must be turned off when you leave the laboratory and put back correctly in the charging station.

2. Microscopes must be used in accordance with the lab manual procedures or your instructor's directions. Return microscopes at the end of lab to the cabinet according to the posted guidelines.

3. Microscope **slides must be wiped clean of oil and must be put back on the right tray or slidebox** according to their numbered code. **Use lens paper ONLY to wipe slides**.

Protective Equipment

1. Students should wear safety goggles in the laboratory when handling chemicals or dissecting specimens. Return goggles to the cabinet after your laboratory session.

2. Chemical operations and specimen dissections should be carried out below the eye level.

3. Eyewash stations are located at the front and/or rear of the laboratory. If chemicals are splashed in eyes, **DO NOT RUB EYES**. Go immediately to an eyewash station and press the lever to start the fountain. Flush eye(s) for at least 15 minutes.

4. All eyewash units in the laboratories also function as a drench hose. Remove contaminated clothing and wash off chemicals splashed or spilled on your skin or body immediately and for 15 minutes.

5. It is recommended that shorts or dresses not be worn in any laboratory. Arms and legs should be covered. Students should wear shoes that cover the whole foot-no sandals.

6. Long and/or loose flowing hair should be secured behind the head and shoulders.

Emergency Incidents

Report any of the following incidents to your instructor immediately and follow the instructor's directions.

a. Personal Injuries

b. Chemical Spills

c. Broken Slides and Glassware

d. Fire

Mott Community College
BIOLOGICAL SCIENCES
Student Laboratory Policies and Procedures

I, ____, have read and understand the policies and procedures for laboratory use and behavior in the Biology Area of Mott Community College and agree to abide by them.

_______________________________   _______________________________

(Student's Signature)   **(Date)**

_______________________________   _______________________________

(Course Code and Title)   **(Instructor's Name)**

Note: Before any student is allowed to work in the laboratory, the student must sign this form to indicate that he or she will comply with the rules for the laboratory. Signed forms must be submitted to the instructor. This form is retained by the instructor and a copy is provided to the Science and Mathematics Division Office.

Laboratory 1
Microscope

Introduction

The microscope is the instrument most frequently used to magnify cells so they may be seen and studied. There are several types of microscopes: the **stereo dissecting microscope** (used to see the detail of large structures), the **compound light microscope** (used to see normal cell structures) and the **electron microscope** (used to see sub-cellular structures).

To begin, you need to become familiar with the microscope and its structures. Although the microscope appears to be sturdy, it is a very precise and delicate instrument. For those of you who have never used a microscope, READ this section very carefully. For those of you already familiar with microscopes, READ through this section for a review. Proficient use of the microscope will be required for several of the laboratory exercises. Students who are unable to use the microscope will be at a disadvantage.

Objectives

1. To develop the ability to use the microscope as a tool in the study of biology.
2. To be able to locate and identify the different parts of the compound light microscope.
3. To be able to find and bring into focus under low and high power various objects on permanent slides and wet mount slides.
4. To learn how to make acceptable wet mounts of various materials (organisms).

Materials

1. Compound light microscopes
2. Lens paper
3. Clean microscope slides
4. Cover slips

5. Plastic droppers

6. Prepared letter (e) slides

7. Tooth picks

8. Methyl blue

9. Pond water with many different micro-organisms

Procedure A: The Compound Light Microscope

First, locate all of the parts of the microscope indicated in figure 1.1. Although microscopes may differ somewhat in appearance, depending on the brand, all are essentially the same. Magnification is accomplished by two sets of lenses. The first set of lenses is the **OCULARS**. These are the black eyepieces at the top of the microscope. These have a magnification of 10x. A **POINTER** is fixed within one of the oculars. In order to use the pointer the ocular can be rotated so the pointer is pointing to the specimen or the specimen can be moved on the **MECHANICAL STAGE** so that the pointer is pointing to the specimen. The other lenses are the **OBJECTIVES**, which are mounted on the **NOSEPIECE**. Our microscopes have three different magnifications: **LOW POWER** is 10x; **HIGH POWER** is 45x; and the **OIL IMMERSION** is100x. To calculate the total magnification when the microscope is used properly, multiply the power of the specific objective by the power of the ocular. Our microscopes have a total magnification of 100x, 450x and 1000x respectively.

The **LIGHT SOURCE** of most microscopes is from a built-in lamp in the **BASE**. The light source is used to concentrate the light into the objective lenses. The power switch is adjustable and can be used to increase the intensity of the light. However, it is recommended that you start with a low intensity of light; you can increase the intensity if it is needed.

Between the light source and the bottom of the mechanical stage are two other important components of the microscope: the **CONDENSER** and the **IRIS DIAPHRAGM**. The condenser is a lens that concentrates the light source through the specimen on the stage. The condenser is similar to an up-side-down funnel for light, which will intensify the light. The iris diaphragm is an adjustable lever to control the amount of light. Initially, the iris diaphragm should be wide open to allow the maximum amount of light to pass through. This is often too bright and the objects may be seen more clearly by closing down the iris diaphragm to reduce the amount of light on the specimen.

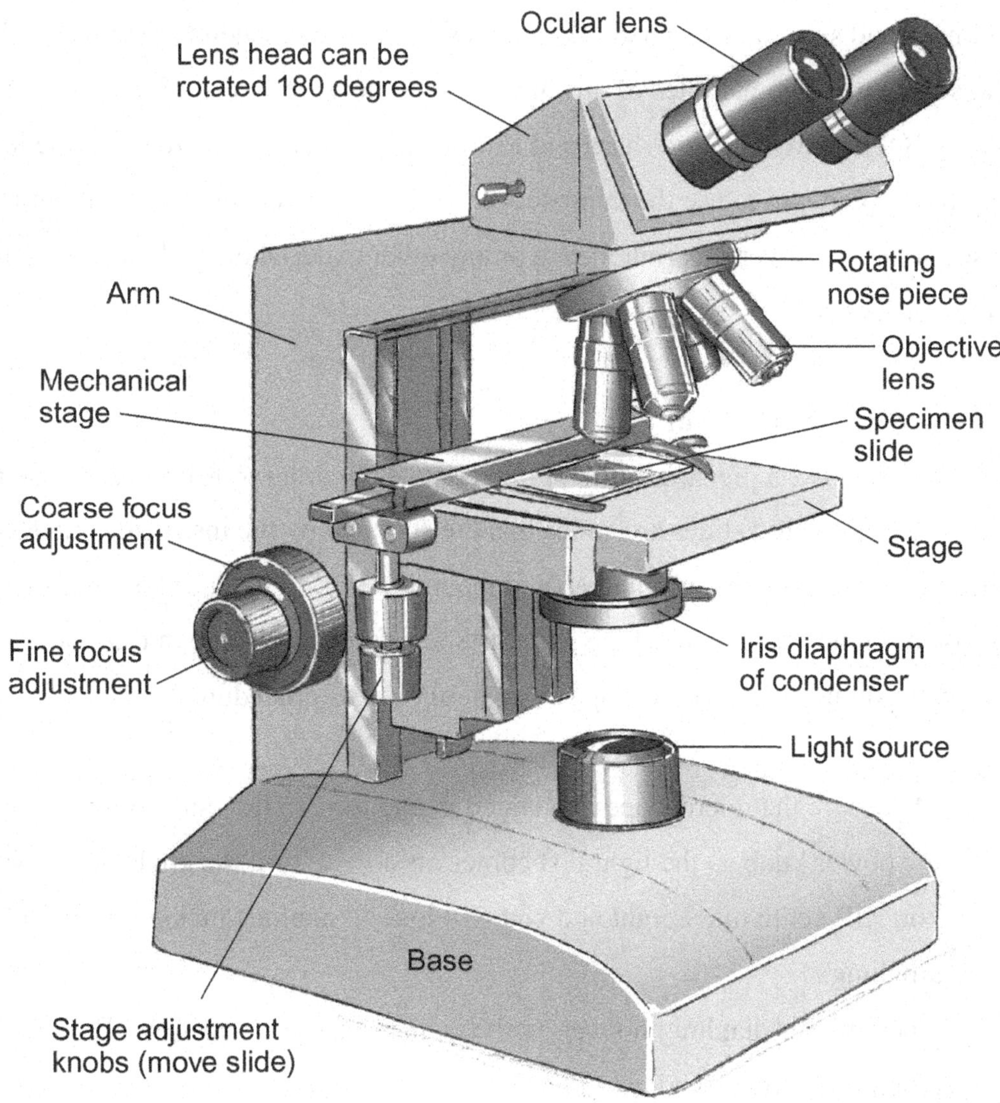

© Image courtesy of Kendall Hunt

Figure 1.1. The Compound Light Microscope

The **COURSE FOCUSING ADJUSTMENT KNOB** and the **FINE FOCUSING ADJUSTMENT KNOB** enable you to move the objective lenses closer to or further from the specimen you are viewing on the stage. Movement of these knobs will bring the specimen into focus.

Over the course of the week, many people use these microscopes. As a result the lenses pick up dust, oil, and other things that will hinder the proper viewing of the specimens and should be cleaned. Sometimes it is obvious when the lenses are dirty because you may see particles of dirt in the lenses, however, most of the time it is not obvious. Objects will be less sharp and less clear. Use lens tissue to clean the lenses every time you begin to use the microscope. If the instructor forgets to put out lens tissue please ask for it (Never use your sleeve or any other material other than lens tissue to clean your lenses).

Procedure B: Finding Your "e"

We will use a prepared slide of the letter "e" in order to learn how to use the microscope. Prepared slides are slides that are supplied by the instructor. Latter in this lab we will learn how to make slides (wet mounts). Before you use the microscope, make sure you are familiar with all of the structures and functions of the microscope. Whenever you using the microscope always follow this procedure:

1. Make sure the microscope is plugged in and turn on the light source. Only turn the power knob so the light just comes on. If you turn the knob all the way to ten you will get to much light and you will lose all contrast making it difficult to see anything.
2. Open the iris diaphragm lever on the condenser about half way. If there is too much light you can adjust this latter.
3. Place your "e" slide on the mechanical stage so that the specimen is in the center of the light that comes through the hole in the middle of the stage.
4. Rotate the objective lenses so that the low power objective lens is directly over the specimen. Make sure the low power objective lens is locked in place it will click. If it does not click into place you will not see anything.

5. Turn the coarse focusing knob so that the objective lenses are at their closes position to the specimen.

6. Look into the oculars; you may have to adjust them to the proper width for your eyes, and rotate the coarse adjustment knob so that the objective lenses rise, causing your specimen to come into focus.

7. Once you have focused the best you can with the coarse focusing knob, you can fine-tune the focus with the fine focusing knob.

8. Next you need to move the specimen that you are looking for into the center of the field of view.

9. If you cannot find what you are supposed to then you need to scan around the field of view until you locate the specimen.

10. Once you have found what you are looking for, you moved it into the center of the field of view and you have focused the best you can using the coarse focusing knob and the fine focusing knob, you can switch to the high powered objective lens.

11. You do this by rotating the objective lenses so that the high-powered objective lens clicks into place. Do not adjust the focusing knobs. It may appear that the high-powered objective is going to touch the slide, but it will not.

12. After you have positioned the high-powered objective, you can sharpen the focus. You do this by adjusting the fine focusing knob **<u>(NEVER THE COARSE FOCUSING KNOB).</u>**

13. If you are unable to find the specimen, you will have to switch back to low power, find what you are looking for and make sure it is in the center of your field of view. Then you can try again on high-power.

14. During this entire process you need to use both of your eyes, that is why there is an ocular for both eyes.   You also need to remember to blink; Staring into the bright light will give you a headache.

Procedure C: Find Organisms in Pond Water

In order to do this we have to make a **<u>Wet Mount</u>**. To make a wet mount:

1. Place one drop of water from the pond water container on a clean glass slide.
2. Next, you need to put a cover slip on the drop of water. You do this by placing the cover slip so that one side of it is touching the water and then let the cover slip fall down into the water.
3. If you put the cover slip on this way, you will not get as many air bubbles. If you get large air bubbles, you can tap the cover slip lightly and work the air bubbles out to one side.

The purpose of this experiment is to get comfortable using the microscope, not to identify all of the organisms. Try and find as many different organisms as you can. Once you find an organism switch to high-power and practice focusing on high-power. This is intended to be practice, to get use to using the microscope.

Procedure D: Cheek Cells

In this procedure, you are going to make a wet mount of the cells from the inside of your cheek. In order to do this you will follow the procedure outlined in Procedure C for making a wet mount. However, you need to put a drop of clean water on the slide and then place the cheek cells into the water. You do this by **lightly** scrapping the inside of your cheek with a toothpick. Once you have done that, place the end of the toothpick into the drop of water on the slide to wash the cells off and then put the cover slip into position.

The cells from your cheek are translucent in color, so you are going to have to stain them in order to see them. You do this by adding one drop of Methyl Blue stain to the edge of the cover slip. Next, take a paper towel and touch the opposite edge of the cover slip with the paper towel. This will wick water up into the towel and draw the stain under the cover, thus, staining the cheek cells.

Laboratory 2

Archaea/Bacteria and Protista

Introduction

The organisms that belong to the kingdoms Archaea, Bacteria and Protista are all considered microorganisms. All of these kingdoms represent the simplest and most primitive of organisms. However, there are vast differences between these Kingdoms; both cellular and morphological differences.

The bacterial kingdoms, Archaea and Bacteria are single celled organisms that are relatively simple. These cells are considered prokaryotic which means they have no nucleus and no membrane bound organelles. They have no nucleus so their DNA is simply suspended within the cytoplasm. Their DNA consists of a single ring structure along with a few extra rings called plasmids. Membranes do not attach the organelles that are present in bacteria. They are simply suspended within the cytoplasm. Two examples of organelles are ribosomes, which are for protein synthesis and are present in all bacteria, and chloroplasts, which is for photosynthesis and present in some bacteria.

Bacteria are exceedingly small, it has been estimated that the weight of living bacteria on Earth exceeds the total weight of all other living organisms by a factor of ten. They are a very diverse group. Bacteria exist in almost every available niche. Not only are the bacteria found in every terrestrial habitat but they are also found in every aquatic habitat, as well as in the air we breathe. Bacteria are also found within and on your body. They are found in saliva, small intestines and your colon.

We usually think of bacteria as being harmful, and some are. However, most of them are not harmful and we depend upon many species for our existence and well-being. If it were not for the nitrogen fixing bacteria plants would not have a source of nitrates and animals would not have a source of food.

Protista are mostly single celled organisms, however, there are some multicellular forms. One way to define this kingdom is to compare them to the bacteria. The kingdom Protista has cells that are eukaryotic. This means the cells have a true nucleus containing their DNA and these cells contain many organelles.

Not all authorities agree on what organisms should be in this kingdom. Some Protistans resemble members of other kingdoms and certain authors tend to put them in other kingdoms. We will look at them using the same classification as our textbook.

We are going to divide the Protista kingdom into the Plantlike Protists, (Algae) the Animal-like Protists (Protozoa), and the Fungus-like Protists. The plantlike Protists are Protists that contain cellulose in their cell walls and have chlorophyll to perform photosynthesis. These Protists are can be both multicellular and single-celled and live in a wide range of aquatic habitats, both marine and freshwater. The animal-like Protists are called protozoa, which means "first animal". These are all single-celled and heterotrophic. They do not have cell wall nor do they do photosynthesis. The different groups of protozoa are classified by their method of locomotion. The fungus-like Protists are mostly multicellular or they are one large cell with many nuclei and organelles but no cell divisions. There are two major groups of fungus-like Protists, the slime molds and the water molds. Both groups are considered decomposers and some individuals are also parasites.

Objectives
1. Recognize the following bacterial shapes: coccus, bacillus and spirillum.
2. Identify the following organisms using the microscope, photographs or preserved specimens: *Volvox sp., Spirogyra sp., Cladaphora sp.,* Diatoms, *Amoeba sp., Paramecium sp., Euglena sp., Radiolaria sp., Oscillatoria sp., Physarum sp.,* Red algae, Brown Algae and *Trypanosoma gambiense.*
3. To be able to recognize whether the following organisms are single-celled to multicellular: *Volvox sp., Spirogyra sp., Cladaphora sp.,* Diatoms, *Amoeba sp., Paramecium sp., Euglena sp., Radiolaria sp., Oscillatoria sp., Physarum sp.,* Red algae, Brown Algae and *Trypanosoma gambiense.*
4. To be able to recognize the asexual and/or sexual reproduction in: *Paramecium sp., Spirogyra sp. and Amoeba sp.*
5. Identify the following organelles: food vacuole, contractile vacuole, plasma membrane, pellicle, chloroplast, nucleus and etc.

Materials

1. Microscope slides
2. Cover slips
3. 1 100 ml beaker
4. Plastic droppers
5. Lens paper
6. Petri dishes with nutrient agar
7. Wax pencil
8. Permanent Monera slides: *Staphyloccous aureus, Bacillus subtilis, Spirillum volutans and Oscillatoria sp.*
9. Permanent Protista slides: *Volvox sp., Spirogyra sp., Spirogyra* conjugation, *Cladaphora sp.*, Diatoms, Diatomaceous Earth, *Amoeba sp., Paramecium sp., Paramecium* fission, *Paramecium* conjugation, *Euglena sp., Radiolaria sp.,* and *Trypanosoma gambiense.*
10. Living specimens: *Oscillatoria sp., Volvox sp., Spirogyra sp., Cladaphora sp.,* Diatoms, *Amoeba sp., Paramecium sp., Physarum sp.* and *Euglena sp.*

Procedure A: Bacteria

1. Observe a living culture of bacteria. In order to do this we will need to collect some bacteria and grow them before there is a large enough colony to be visible. Obtain a petri dish from your instructor. The petri dish will have nutrient agar on one side. This is where the bacteria will grow once you introduce them to the media. The agar is similar to gelatin so you need to be careful not to damage it when you introduce the bacteria.

   In order to introduce the bacteria all you need to do is touch the petri dish side with the agar lightly to something. Simply press the agar side of the petri dish on to a doorknob, bottom of your shoe, money, etc. Make sure not to press too hard or you will damage the agar and the bacteria will grow into the agar and not on top of the agar, making it hard to see the colonies.

   Once you have inoculated the agar with bacteria you need to label your petri dish with your name, date, and lab section. After you label the petri dish, place it in the incubator with the agar side up to prevent the buildup of condensation. We will leave your petri dishes in the incubator until the next laboratory period, which will be plenty of time for the bacteria to reproduce enough so that there is a large colony. These colonies will be large enough to be seen without the microscope.

2. Observe a prepared slide of *Staphylococcus aureus*. This is a common bacterium, which is relatively harmless but can be harmful under the right circumstances. You need to see the shape of these bacteria. These are very small organisms so you will have to use the oil immersion objective lens. Your instructor will demonstrate how to do this. The cell shape is considered a coccus.

    a. What shape are the bacteria that are coccus? _____________________

3. Observe a prepared slide of *Bacillus subtilis*. The genus *Bacillus* is named for the shape of these cells. The shape is called bacillus. These organisms are larger than the *Staphylococcus aureus*, however, you will still have to use the oil immersion objective lens.

    a. What shape are the bacteria that are bacillus? _____________________

4. Observe a prepared slide of *Spirillum volutans*. This bacterium is a common saprophyte, which means it lives on dead and decaying organic material. This particular organism can be common sometimes in the human mouth. This genus is also named for its shape. The shape is called spirillum. These bacteria are quite a bit larger than the previous bacterial cells so you will be able to see them with the high-powered objective lens.

    a. What shape are the bacteria that are spirillum? _____________________

5. Examine both a prepared slide and a wet mount of the *Oscillatoria sp.* This particular bacterium lives as a colony in the shape of a long filament (thread-like). The filaments are made of rectangular cells that are attached together in a long chain. These bacteria are common in stagnant water.

At one time these bacteria were called blue-green algae. Today some microbiologists call these bacteria blue-green bacteria, which is a truer name. The technical name is Cyanobacteria, which means blue-green bacteria.

    a.   Sketch the *Oscillatoria sp.*

Procedure B: Plantlike Protista (Algae)

Chrysophyta

1. These organisms belong to the phylum Chrysophyta, which are the Golden-brown Algae. Although the diatoms contain the same chlorophyll as green plants they also contain large amounts of other pigments, which mask the green, and the organisms appear golden-brown. This is one of the most abundant algae in the ocean. The diatoms have a shell that is made out of silicon dioxide (glass). When these organisms die these glass shells collect at the bottom of lakes and oceans. Examine both a prepared slide and a wet mount of the Diatoms and sketch some of the different forms that you see.

Prepare a wet mount of Diatomaceous Earth. Examine your slide under the microscope. Diatomaceous Earth is the glass shell remains of dead diatoms. Over the years these organisms die and their shells do not decompose very quickly. The broken glass shells provide an excellent material for an abrasive substance or a filtering agent.

Euglenophyta

2. Prepare a wet mount of *Euglena sp.* Observe them under the microscope. Notice the manner in which the *Euglena* swims. The *Euglena* has a flagellum for locomotion located at one end of the cell. They also have chloroplast to do photosynthesis and synthesize their own food. They can also ingest bacteria and other microbes for food.

Chlorophyta

3. Observe both a prepared slide and a wet mount of the colonies of *Volvox sp.* Examine under low power because these colonies are so large. The *Volvox* colony has a spherical shape. It is a hollow ball with gelatinous material holding the cells together to form the sphere. Each cell has two flagella making the entire colony mobile.

    Locate some *Volvox* colonies with large, densely green spheres within. These are daughter colonies. They arise when some of the cells on the surface of the colony reproduce at a more rapid rate and produce an invagination that develops into a small sphere. When the *Volvox* becomes too large it is more susceptible to physical damage. When this happens, the small spheres or daughter colonies are released. This is a form of asexual reproduction. The *Volvox* also produces gametes and reproduces sexually. Figure 2.1.

4. Prepare a wet mount of *Spirogyra* sp. and compare it to a preserved specimen. Examine under low power on the microscope. This is an unbranched filament with long, rectangular cells that are joined end to end to form a thread-like filament. Many filaments become densely tangled to form huge mats of *Spirogyra*, which float upon the surface of ponds and lakes. Examine one of the *Spirogyra* cells under high-power. The most striking feature is the spiral, green chloroplast within the cell. The function of the chloroplast is to do photosynthesis. Some *Spirogyra* has extra chloroplast not in the spiral shape.

5. On the prepared slide locate the *Spirogyra* that is conjugating. This is a form of sexual reproduction. Compare the conjugation on the slide with Figure 2.2. Two filaments of *Spirogyra* align side by side. Small projections grow from the cells of each filament towards cells of the opposing filament. When they meet a tube is formed, called a conjugation tube, between each pair of cells. The content of each cell compacts into a small sphere. The cellular contents of each cell of one of the filaments moves through the conjugation tube to the adjacent cells of the other filament and the materials fuse resulting in a zygote. When the process is completed, the cells of one filament will contain zygotes while the cells of the other filament will be empty and dead. The zygotes will then be released and each zygote is capable of growing into a new filament.

   a. Is this type of reproduction sexual or asexual? ______________________

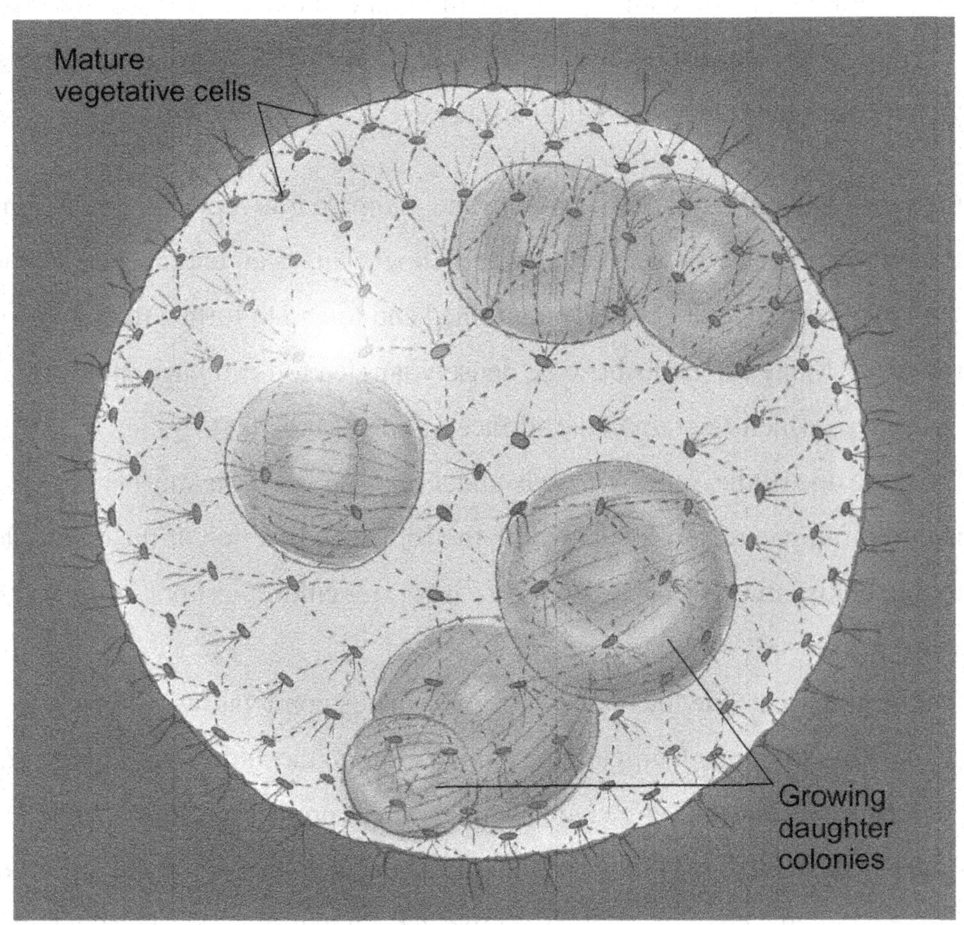

© Image courtesy of Kendall Hunt

Figure 2.1. *Volvox sp.*

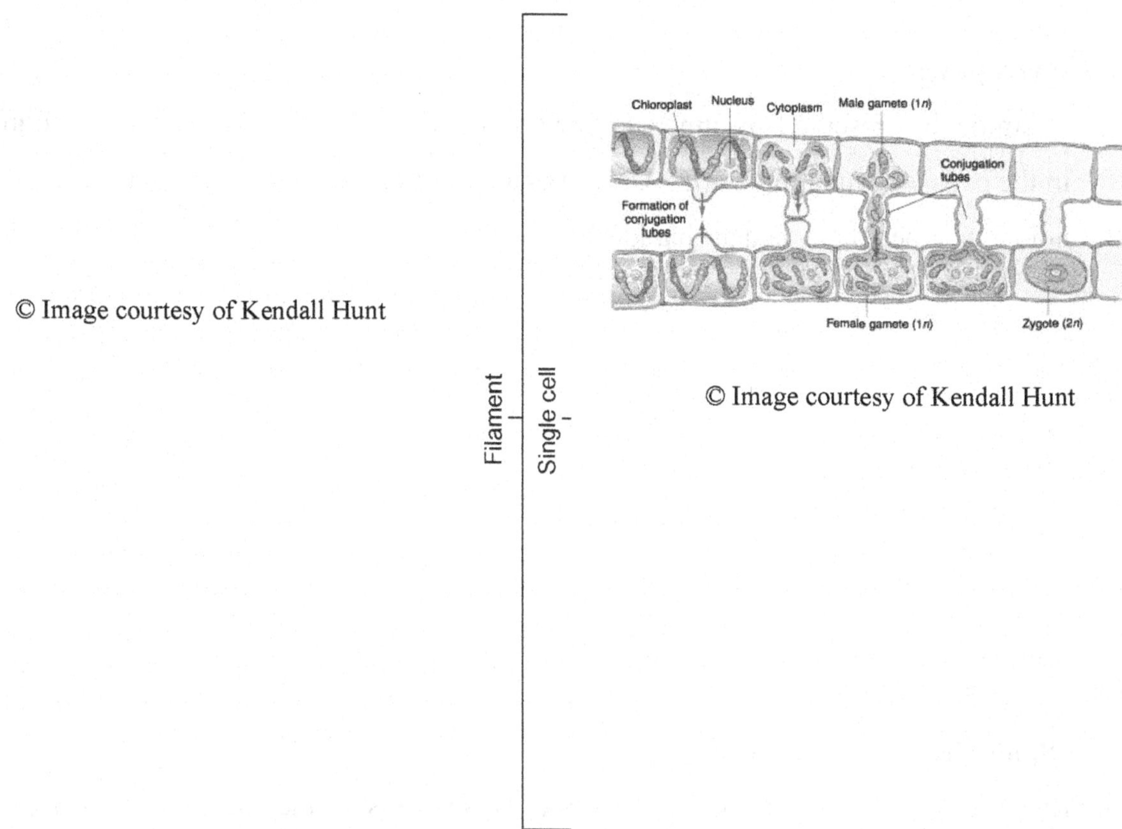

© Image courtesy of Kendall Hunt

© Image courtesy of Kendall Hunt

Figure 2.2. *Spirogyra sp.*, conjugation

6. Prepare a wet mount of *Cladophora* sp. Examine this under the microscope and sketch. The *Cladophora* is a fresh water alga, found in lakes, streams and ponds. Under low power you may be able to note the resemblance to *Spirogyra*. However, it differs from *Spirogyra* in many ways. If you examine several filaments you will find some of the filaments are branched. Occasionally you may find that one end of the filament has many branches. This is called the holdfast. The holdfast attaches the *Cladophora* to rocks and other objects in the water to provide stability for the filaments.

Phaeophyta

Examine the preserved specimens of the Brown Algae (kelp). These algae are found in the oceans around the world. Draw a diagram of one individual and label the holdfast, stipe, blade, and air bladder.

Rhodophyta

Examine a preserved specimen of the Red Algae. Most of the red algae are marine and found in tropical oceans. Note the red pigmentation.

Procedure C: Animal-like Protista (Protozoa)

Mastigina (Flagellates)

1. Examine a prepared slide of *Trypanasoma gambiense*. The *Trypanasoma* is one example of protozoa that uses a flagellum for movement. This slide is of a human blood smear. These organisms are very small and you will have to use high-power to see them. These are actually smaller than the red blood cells. This particular organism causes African sleeping sickness. It is transmitted to humans by the tsetse fly. Figure 2.3.

Sarcodina (Amoeboid)

2. Examine both a prepared slide and a wet mount of the *Amoeba sp.* In order to make good wet mounts remove some of the material from the <u>bottom</u> of the culture of *Amoeba*. These organisms are large enough to be seen under low power. It may be difficult to differentiate the *Amoeba* from other organic debris in the culture. Watch the *Amoeba* for a while and notice the way it moves. The extensions of the *Amoeba* are called pseudopodia, which mean false foot. Figure 2.4.

    If your *Amoeba* is alive you will be able to observe the cytoplasm moving. This activity is called cytoplasmic streaming. The function of cytoplasmic streaming is to provide circulation within the *Amoeba* and to aid in movement of the *Amoeba*. The *Amoeba* reproduces asexually by fission. Fission is simply the dividing of one organism to form two identical daughter organisms.

Ciliophora (Ciliates)

3. Prepare a wet mount and compare the living *Paramecium* with the preserved *Paramecium* on a prepared slide. *Parameciums* are large and swim fast so you must have patience when looking for them. Figure 2.5. Smaller protozoa may also be present. These are not *Paramecium*. These provide food for the *Paramecium*. Observe the response of the *Paramecium* when it encounters objects in the water.

a. Describe how the *Paramecium* reacts when it encounters an object in the water. ____

4. Examine a prepared slide of fission in the *Paramecium*. This is asexual reproduction. When the *Paramecium* goes through fission it divides into two identical daughter cells just like the *Amoeba*. Protozoa that reproduce asexually are typically found in habitats that are stable and not changing. If the parent organism can survive in the stable environment then their identical off spring should be able to survive in the same stable environment. Sketch the *Paramecium* going through fission.

5. Examine a prepared slide of the *Paramecium* undergoing conjugation. Conjugation is a type of sexual reproduction. The *Paramecium* pair off and each one exchanges some genetic material with its mate. The new genetic material fuses with the old to form new combinations of genes. At this time each of the *Paramecium* divides producing offspring that are genetically different from one another. Protozoa that reproduce sexually are typically found in environments that are not constant and are unstable. If the offspring are genetically different from one another and the environment changes, there is a better chance that some of the offspring will survive.

a. Sketch the *Paramecium* going through conjugation.

Procedure C: Fungus–like Protista

Myxonycota

1. Observe the living Slime Mold (*Physarum*). Slime molds are unusual organisms. At different times in their life cycle they will behave like animals and at other times they behave like plants. They behave like animals because they move around and consume food. They behave like plants because they produce a stalk to release spores. Some biologists consider the slime mold to be a fungus primarily because they are decomposers. However, they are considered a Protista. Sketch the slime mold.

Figure 2.3 *Trypanasoma gambiensa*

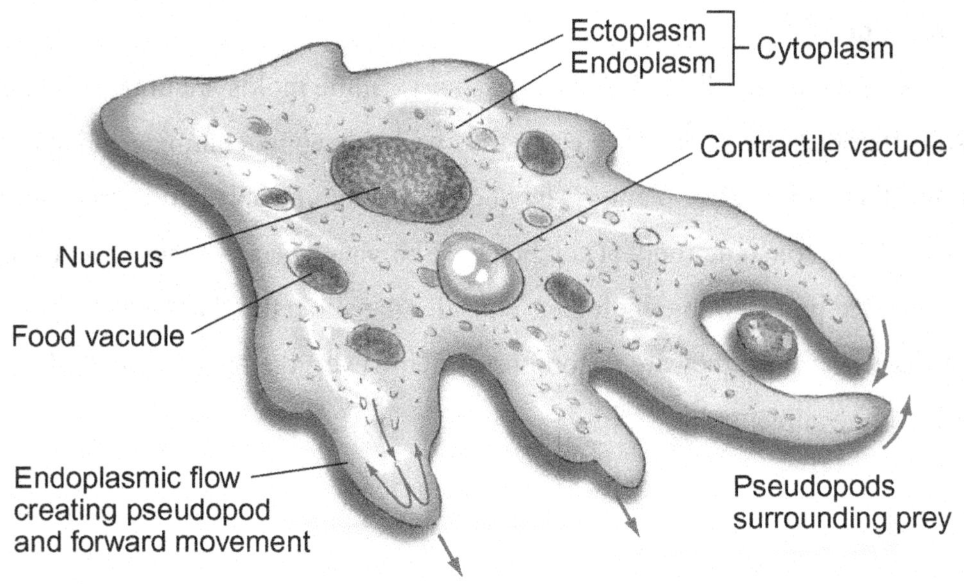

Figure 2.4 *Amoeba sp.*

© Image courtesy of Kendall Hunt

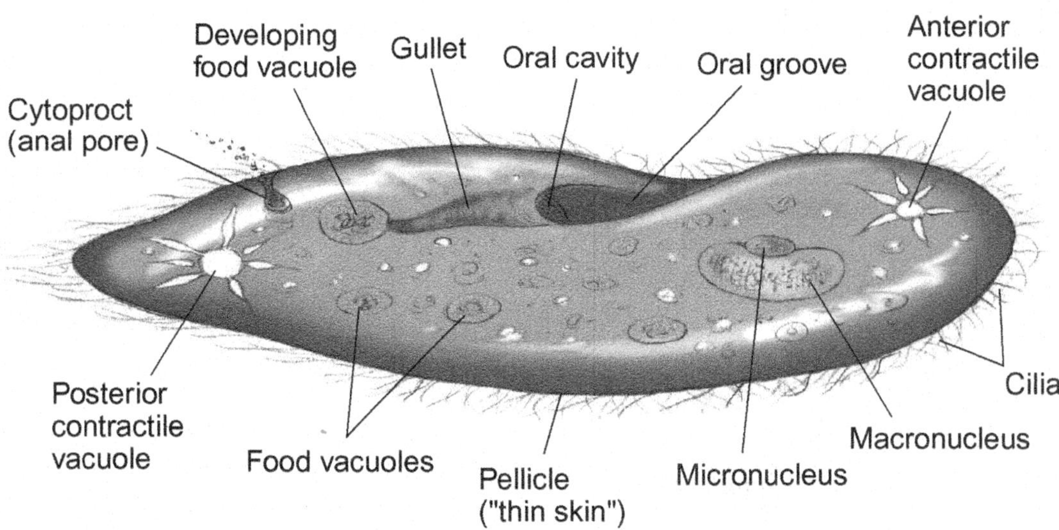

Figure 2.5 *Paramecium sp.*

© Image courtesy of Kendall Hunt

Laboratory 3

Fungi

Introduction

Fungi along with bacteria are the decomposers of our world. For the most part, the fungi's niche is as a saprophyte, to break down dead bodies. Through this process the fungi releases nutrients that have been stored away in the dead body's tissue. These nutrients are released into the environment and are now available for plants and other organisms to use. Some fungi are considered parasites. These fungi decompose living tissue. Some examples of these parasitic fungi are ring worm and athletes foot. The fungi that we are going to look at are saprophytes or they are parasites on other organisms, not us.

The fungi's body is divided into two stages: the fungal body which is persistent and the fruiting body, which is only present when spores are going to be released. The fungal body is made up of hyphae. Hyphae are masses of cobweb-like material that grows quite extensively on a food source. If the food source is plentiful the fungal body is capable of covering the food source with hyphae so that it appears to be hairy. An example of this is downy mildew of grapes. The fruiting body is a reproductive structure sometimes called a sporangium. The fruiting body is a stalked structure that produces the spores. This allows the spores to be dispersed by the wind when they are mature. Once these spores are released the fruiting body disappears.

Extracellular-enzymatic digestion and absorption is how one would describe the fungi's digestive system. As the fungi grow across its food source, the fungi secrete enzymes out of their body onto the dead organism. These enzymes begin to degrade or breakdown the dead organism. As the dead organisms nutrients are released from its tissue, the fungus absorbs the nutrients. However, the fungus is not very efficient at absorbing the nutrients once they have been broken down and most of the nutrients are lost to the environment. These nutrients that have been lost to the environment are what are available for the plants and other organisms to absorb and use.

Objectives

1. To become aware of the ecological role that fungi play in our environment.
2. To have a better understanding of what a mushroom is and what it is not.
3. To be able to explain what the difference is between the Fungal Body and the Fruiting Body.
4. Recognize the following fungal structures: Hyphae, Sporangia, Zygospore, Ascospore, Basidiospore, Ascus, Basidium, Stipe, Cap, Annulus, and Gills.
5. Identify by using a microscope or other means the following: *Rhizopus sp., Peziza sp.,* Corn Smut, Puffball, *Coprinus sp.,* Bracket Fungi, Coral Fungi, Mushrooms, Yeast, Powdery Mildew, *Penicillium sp.* and the Lichens (Fruticose, Foliose, and Crustose).

Materials

1. Microscope slides
2. Cover slips
3. 1 100 ml beaker
4. Plastic droppers
5. Lens paper
6. Living cultures of yeast and mushrooms
7. Prepared slides of: *Rhizopus sp., Rhizopus* Conjugation, *Peziza sp.,* Powdery Mildew, *Penicillium sp.,* and *Coprinus sp.*
8. Preserved specimens of: Leaves with Powdery Mildew, Morels, Coral Fungi, Bracket Fungi, Puffballs, Corn Smut, and Lichens (Fruticose, Foliose, and Crustose).

Procedure A: Zygomycota (Algal-like Fungi)

The algal-like fungi were named this way because of their resemblance to algae. They resemble algae in their appearance and in the way they reproduce. They differ from the rest of the fungi by lacking partitions between individual cells of the hyphae. The particular algal-like fungus that we will look at will be *Rhizopus sp. Rhizopus* is our typical bread mold. However, it is not the only bread mold.

1. Examine a living culture of *Rhizopus* under a dissecting microscope. *Rhizopus* is composed of many interlocking hyphae. You may also be able to locate little round black balls. These are the fruiting bodies (sporangium). They are filled hundreds of spores that are ready to be released and inoculate more bread. Sketch what you see.

2. Examine a microscope slide of the sexual reproduction of *Rhizopus*; this is called conjugation. Figure 3.1. *Rhizopus* has two types of dissimilar spores. One is considered positive and the other is considered negative. If these two dissimilar spores land in close proximity to one another there is a chance for sexual reproduction. This occurs by the two spores growing hyphae. As the two different hyphae grow close together each hyphae will grow small extensions towards each other. When these extensions meet they form a large mass of cells called a zygospore. Out of the zygospore grows the sporangium, which when mature, contains spores. The final step in the cycle is dispersal of the mature spores.

Procedure B: Ascomycota (Sac Fungi)

The sac fungi differ from the algal-like fungi in two basic ways. First, the sac fungi have hyphae with distinct cells. Each cell has a separate nucleus. The second difference is a structural difference during sexual reproduction. This fungus produces a sac-like structure called an ascus. Each of these asci contains eight ascospores. This is a large group of fungus with considerable impact on people both good and bad. We will only consider a few within this large group.

1. Examine the preserved specimens *Peziza*. These cup-like structures are found emerging from decaying logs in humid forests. The inner surface of the cup is lined with asci containing spores.

2. Examine a prepared slide of *Peziza*. Locate the long tube-like asci. Each ascus contains eight ascospores. When mature the asci break open and release the spores. The sac fungi also have dissimilar spores, positive and negative. These spores also germinate and grow hyphae. However, when these dissimilar hyphae come in close proximity to one another they fuse. This fusion produces binucleated hyphae. The next step in the production of spores is for the fungus to produce a fruiting body stemming from the binucleated hyphae. In this fruiting body is where the ascus and the eight ascospores are found. The final step in the production of spores is when the spores are released causing the cycle to repeat. Figure 3.2.

3. Examine a prepared slide of powdery mildew. Powdery mildew grows on the surface of leaves, lilacs, phlox, etc. The hyphae produce a huge mass that can cover the entire surface of the leaves. This can make the leaves appear gray and fuzzy. The powdery mildew is considered a parasite because they actually penetrate into the leaves and take nutrients from the leaves. As you observe the powdery mildew notice the black spherical structures with hyphae growing out from them. These are the fruiting bodies of the powdery mildew. Inside of these structures are the asci with the ascospores. Figure 3.2.

4. Prepare a wet mount of yeast. Yeast is an unusual fungi because it is a single celled individual. Yeast is also different because it does not reproduce sexually it

only reproduces asexually. The yeast does this by budding. This means that the original cell simply grows another cell off its side.

In order to see this you must have patience. When you put your slide under the microscope the yeast become activated and begins to release CO_2. There is so much CO_2 being released that the yeast cells appear to be moving under their own power. They are not of course. They are, however, being pushed by the CO_2 that is being released. Once the yeast cells heat up from the light of the microscope they will stop moving. When this happens, find an area where you can observe individual yeast cells under high-power. Scan around until you find what appears to be one large cell with one little cell stuck to it. This will be a budding yeast cell. When the new bud becomes the size of the parent cell, the two will separate. Sketch a budding yeast cell. Figure 3.2.

5. Examine the fruiting body of a morel. Some people like to call them a morel mushroom. Technically they are fruiting bodies not mushrooms. These are one of the few delicacies that Michigan has growing naturally. The best time of the year to find them is around Mother's Day. The best sites to find them are near dead elm trees or dead apple trees in the forest. However, if you are not sure if it is a true morel, DON'T EAT IT!

6. Examine a prepared cell of *Penicillium sp*. *Penicillium* is a fungus that we use as an antibiotic. It kills bacteria. *Penicillium* is a small fungus and you are going to have to use high-power to see it. The fruiting bodies appear as little witches brooms, with a long stalk and the ascospores clustered on one end. Sketch the *Penicillium* fruiting body.

Procedure C: Basidiomycota (Club Fungi)

The club fungi have hyphae like the sac fungi. Each cell is separated. However, they differ from the sac fungi in their reproduction. The club fungi do not produce eight ascospores in an ascus. Instead they produce four basidiospores on a club-like protrusion called a basidium that is found on the gills of a mushroom. The mushroom is the fruiting body for this group of fungus.

1. Examine both the model of the mushroom and the real specimens. The mushroom is the reproduction structure of the club fungi. Locate the following structures. The stem-like structure is called the stipe. On top of the stipe is the cap. Under the cap is where the gills are found and on the gills is where the spores are produced. Covering the gills attaching the cap to the stipe on some mushrooms is the annulus. Usually by the time the spores are mature and ready to be released the annulus has broken down. Figure 3.3.

2. Examine a prepared slide of *Coprinus sp*. *Coprinus* is a fungus, which has a very small mushroom. This slide is a cross-section through the cap showing the gills radiating out from the center. Switch to high-power and find the basidiums with the basidiospores on them. On a mature basidium there are four, chestnut-brown basidiospores. If there are less than four basidiospores, they probably fell off. The life cycle of the club fungi is very similar to the life cycle of the sac fungi. They also have dissimilar spores, positive and negative. These spores also germinate and grow hyphae and when these dissimilar hyphae come in close proximity to one another they fuse. This fusion also produces binucleated hyphae. The next step is for the fungus to produce a fruiting body, a mushroom, stemming from the binucleated hyphae. Figure 3.3.

3. Examine the fruiting body specimens of the bracket or shelf fungus under a dissecting microscope. These fungi reside in trees and these mushrooms grow off of the wood. Examine the undersurface of this mushroom. They do not have gills. Instead they have pores. In these pores is where the spores are found. The

spores are released and enter another tree through cracks and other openings in
the tree. These spores germinate and the hyphae grow into the wood of the tree
causing decay and ultimately death.

4. Examine the preserved specimens of the coral fungi. These are fungi that are
common in some parts of our area. They live in a variety of habitats just as other
mushrooms do.

5. Examine the preserved puffball fungi. They are called puffball because they are
shaped like a ball and when touched they release a puff of spores, like smoke out
of a chimney. Be careful not to release all of the spores when you touch them.

6. Examine the preserved specimens of corn smut. Corn smut is a fungus that
affects the kernels of corn while the ears are still on the plant. The corn smut
basically converts the kernels of corn into masses of spores. This causes the corn
to appear disgustingly nasty. As a result it is inedible.

Procedure D: Mycophycophyta (Lichens)

Lichens are a combination of algae and fungi living together in a symbiotic
relationship. The alga, which normally requires water, is preserved by the fungal partner.
The algae may occur elsewhere without the fungus but the fungus does not occur
elsewhere without the algae. The fungus is the only member of the pair that reproduces
sexually with spores. The fungi, which are symbionts in lichens, are either sac fungi or
club fungi.

1. Examine the preserved specimen of the fruticose lichen. They appear to be
complexly branched. Normally grow on very dry soil or in trees.

2. Examine the preserved specimen of the foliose lichen. The foliose lichen appears
to be exfoliating or peeling off what it grows on. The foliose lichen is usually
found growing on trees, both living and dead.

3. Examine the preserved specimen of the crustose lichen. The crustose lichen appears to be a crust growing closely pressed against what it is growing on. The crustose lichen lives in very inhospitable places such as bare rock and tombstones.

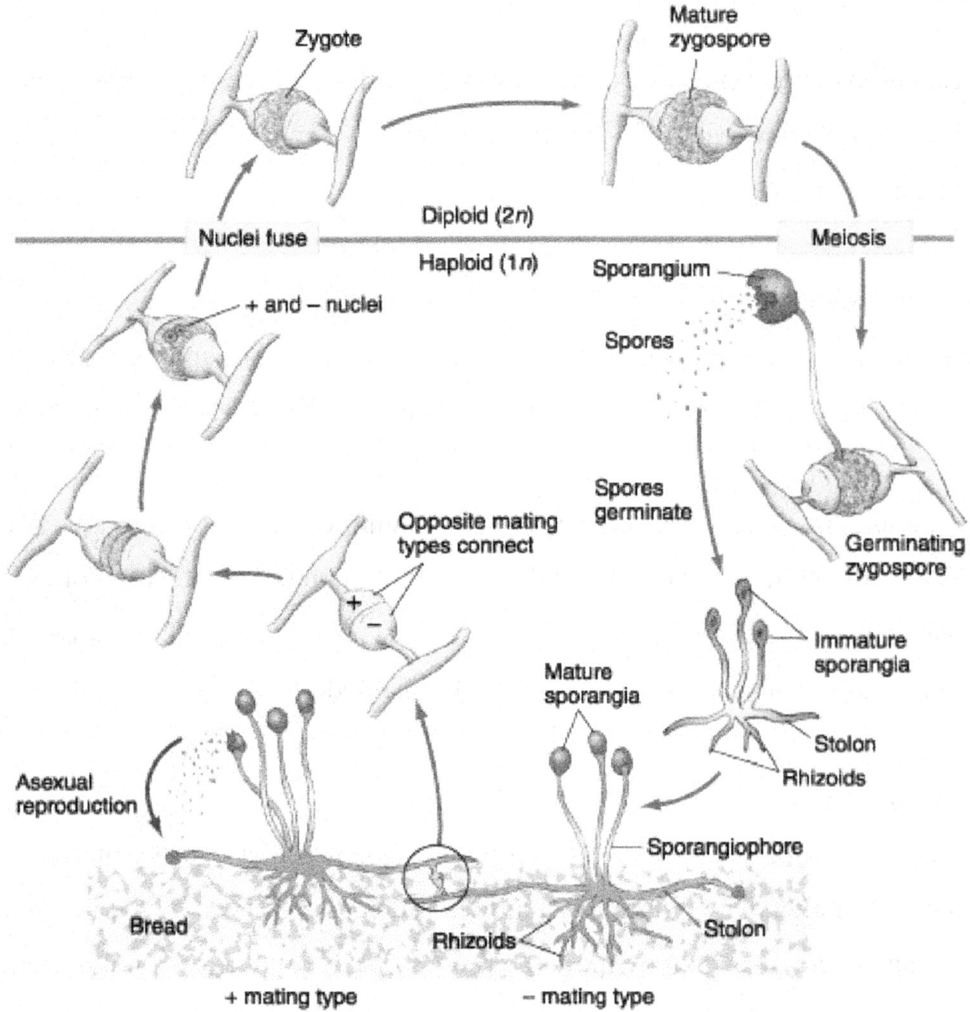

Figure 3.1 *Rhizopus sp.* Life Cycle

© Image courtesy of Kendall Hunt

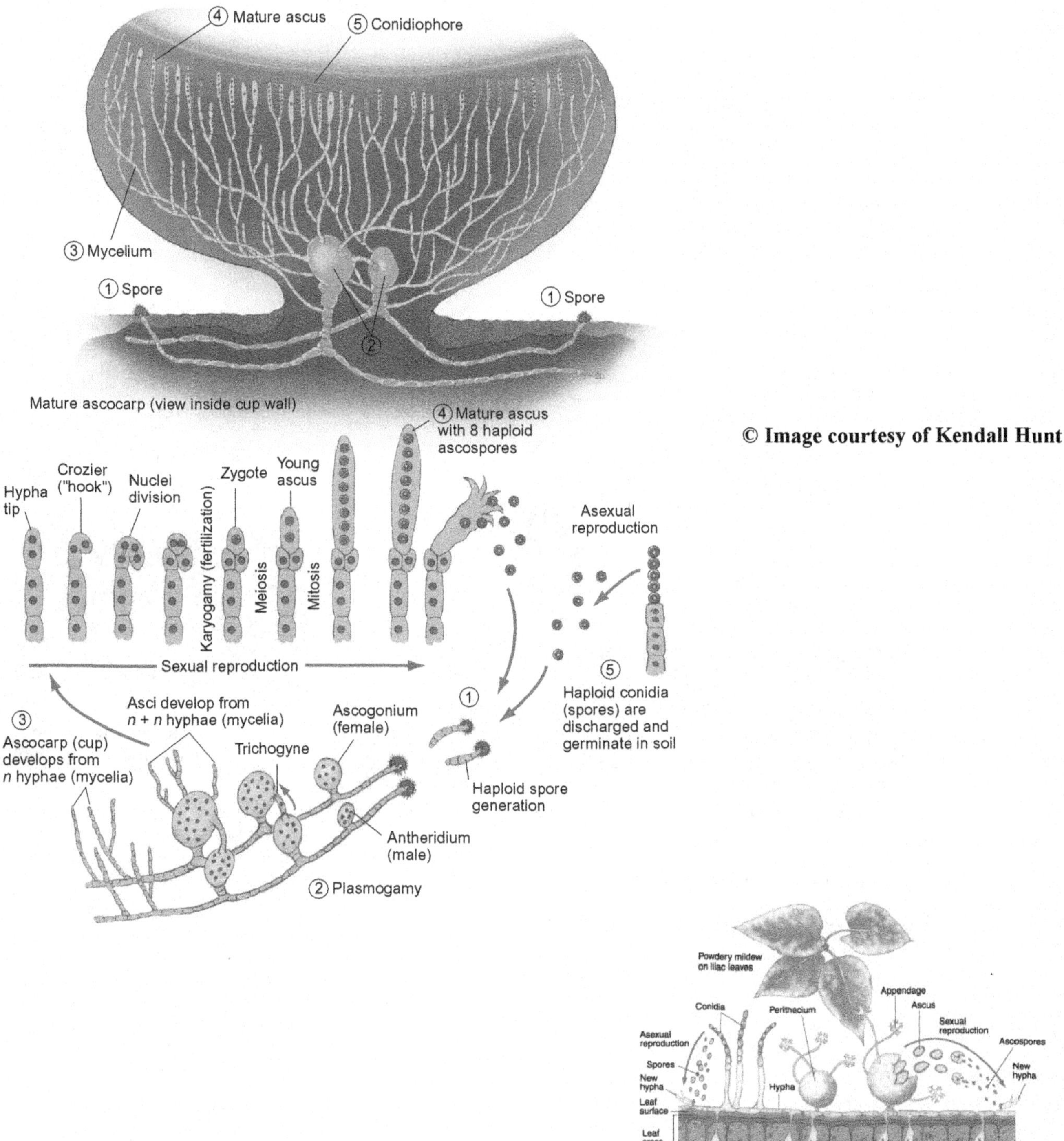

© Image courtesy of Kendall Hunt

Figure 3.2 Types of Ascomycota

© Image courtesy of Kendall Hunt

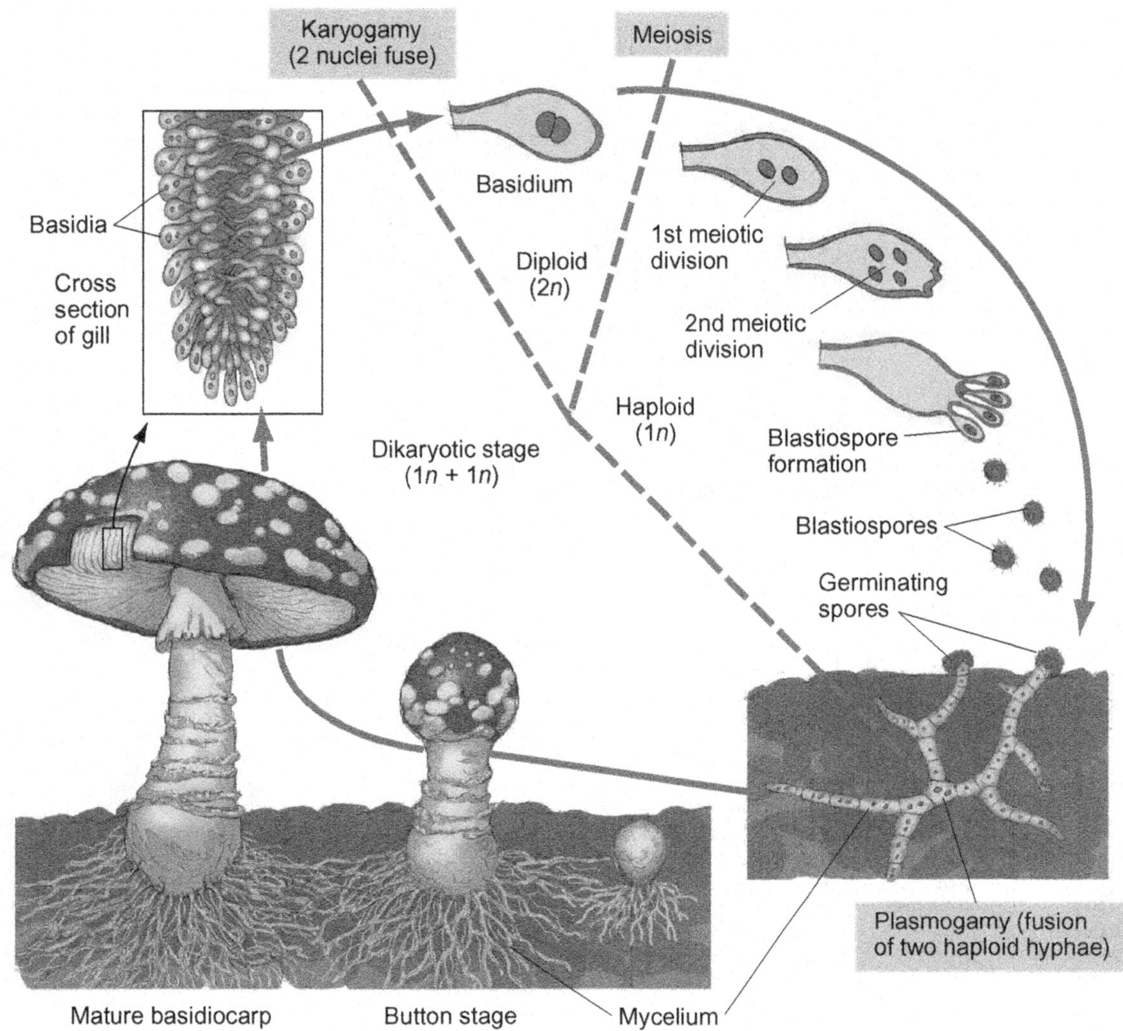

Figure 3.3 Basidiomycota (mushroom)

Laboratory 4

Mosses and Other Cryptic Plants

Introduction

Mosses belong to the phylum Bryophyta of the Plantae kingdom. These plants are very primitive compared to the flowering plants. First, they do not have flowers and do not make seed. Instead, they make spores that blow around when they are mature in order to disburse. Next, they have no vascular tissue. The vascular tissue in plants is an internal transportation system for the movement of water, minerals, and nutrients. Finally, they have no true roots, true stems and true leaves. This is because they have no vascular tissue.

Members of the Plantae kingdom have an alternation of generations. The two generations of the life cycle are the gametophyte and sporophyte generations. The gametophyte generation is the generation that produces the gametes (sex cell). The sporophyte generation produces spores. In the mosses, the gametophyte generation is the dominant generation. The green leaf-like portion is the gametophyte. The sporophyte generation is relatively small and grows on top of the female gametophyte. The sporophyte is also not persistent like the gametophyte.

Objectives

1. To become familiar with the moss and their structures.
2. To be able to distinguish between the sporophyte and gametophyte generation.
3. To be able to recognize the liverwort and other moss relatives.

Materials

1. Living specimens of mosses
2. Living specimens of a liverwort or a hornwort
3. Preserved specimens of a moss with the sporophyte
4. Prepared slides of: moss antheridia, moss archegonia, moss capsule/sporangia and moss protonema
5. Lens paper

Procedure A: Mosses

1. Examine the living specimens of the moss. Determine which part of the moss is the gametophyte. Also, determine if the sporophyte is present. Figure 4.1.

2. Examine a prepared slide of the moss antheridia. Compare what you see with the diagram of the moss life cycle. Figure 4.1. The antheridia are the oval structures at the tip of the male gametophyte. These structures are packed with numerous sperm cells, when they are mature. Each sperm cell has two flagella, which allows them to swim when they are released in order to find and fertilize an egg.

3. Examine a prepared slide of the moss archegonia. Compare what you see with the diagram of the moss life cycle. Figure 4.1. The archegonia are elongated structures that are found at the tip of the female gametophyte. Near the base of the archegonium there will be a single large cell called the egg.   The egg is not released. It remains in the archegonium and is fertilized there. Once fertilization is complete the sporophyte will grow from the original site of the egg.

4. Examine a preserved specimen of a moss with sporophytes. Notice that the sporophyte grows on top of the female gametophyte. At the tip of the mature sporophyte grows the capsule. When the capsule is mature it contains spores. When the spores are mature they are released.

5.   Examine a preserved slide of a moss capsule. The moss capsule is very large when observing it under the microscope. When using low power the entire field of view will be the capsule. Notice that within the capsule there are two chambers. These chambers contain hundreds of spores. Normally, these spores are released from the capsule and disseminated through the air. When the spores land on suitable habitat they germinate and continue the life cycle. Sketch the capsule, be sure to include the two spore chambers. (There is room on the top of the next page for your sketch.)

6. Examine a prepared slide of the moss protonema. These are the algae-like
    filaments that grow from a spore when it germinates. They spread over the
    surface of the substrate and produce small buds. These buds will become the
    male and female gametophytes. Sketch the protonema.

Procedure B: Liverworts and Hornworts

Liverworts appear flat, slimy and dark green. They are usually found in areas that are
very moist but not inundated with water. An example of suitable habitat would be in
cracks of rocks that are kept wet by the mist of a waterfall. Hornworts are similar to
liverworts but differ by producing green horn-like protrusions.

1. Observe the living specimen of the liverwort or hornwort.

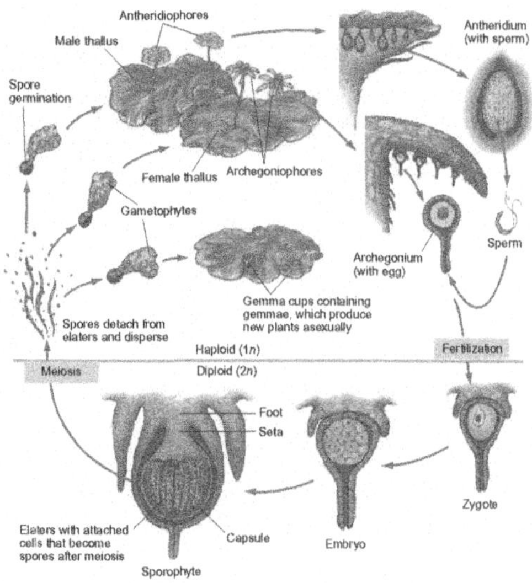

© Image courtesy of Kendall Hunt

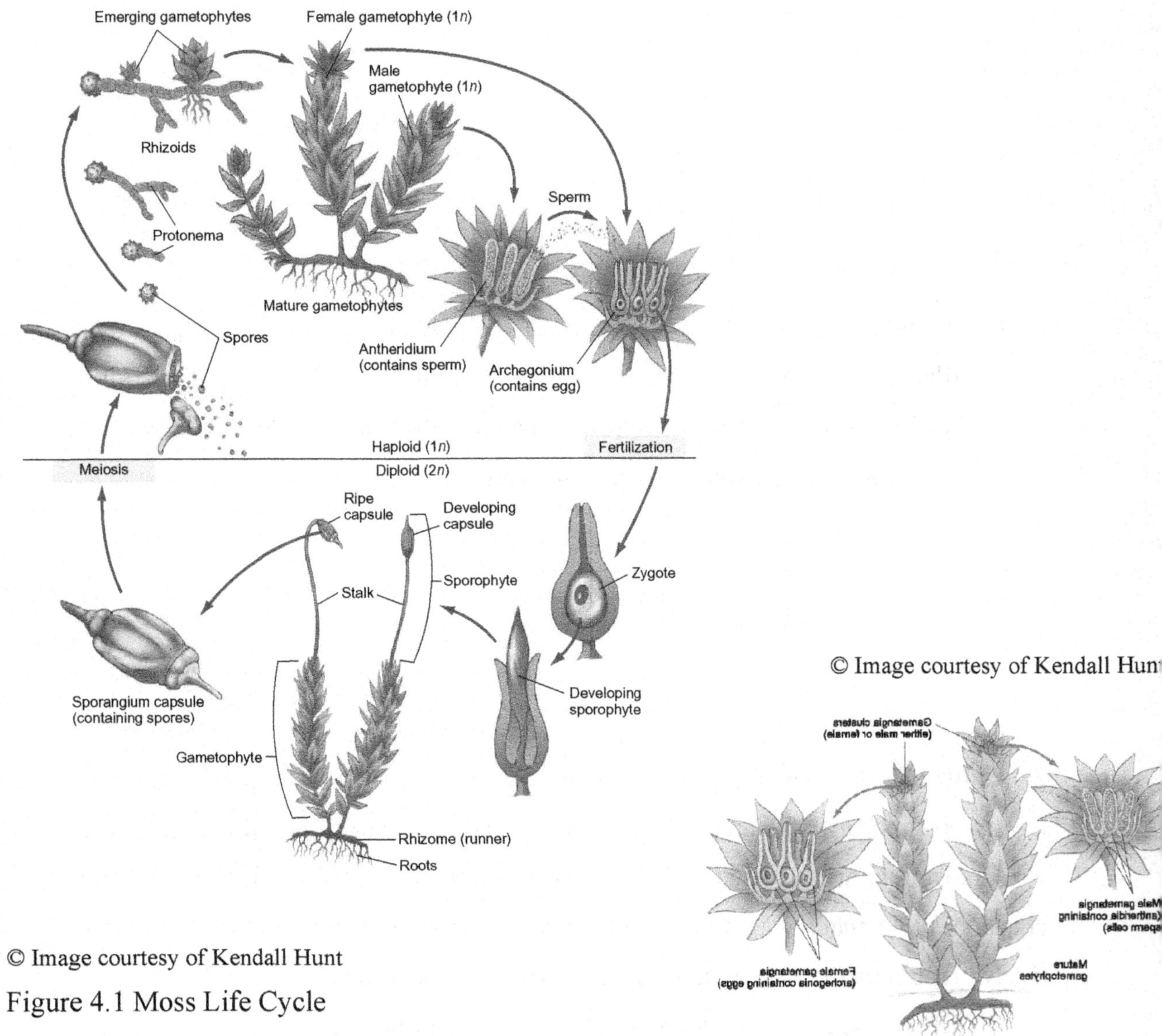

© Image courtesy of Kendall Hunt

Figure 4.1 Moss Life Cycle

© Image courtesy of Kendall Hunt

Laboratory 5

Ferns and Their Relatives

Introduction

The ferns and their relatives are a group of plants that have vascular tissue but do not make seeds. Ferns have vascular tissue, unlike the mosses. This is the tissue that functions as a transportation system throughout the plant. There are two types of vascular tissue, phloem and xylem. The phloem vascular tissue carries the products of photosynthesis from the leaves down throughout the plant. The xylem vascular tissue carries water and minerals that have been absorbed from the ground up throughout the plant. Ferns do not make seeds they make spores, which is similar to mosses. The basic function of the spore is similar to a seed but is very different in structure. The function of the spore is to disseminate the offspring. It is a mechanism, which allows the disbursal of the offspring. Structurally the spore is quite different from the seed. The seed has stored energy within it and allow for it to remain dormant for a time, the spore does not.

Ferns come in many shapes and sizes. In Michigan there are ferns that are as small as two centimeters (*Ophioglossum sp.*) and as big as 1.5 meters (*Mattuchia sp.*). In the tropics there are tree ferns that can be 6 to 12 meters tall. In the past, tree ferns were more widespread and their fossils can even be found in Michigan.

Ferns have very particular names for their structures. The fern leaf is called a frond. As a new frond emerges it essentially unrolls, as it gets bigger. The new frond is called a fiddlehead because of its appearance. Some fronds can be very large and still have a fiddlehead at the end. This would indicate that the frond is still getting larger. Another structure is the rhizome. The rhizome is an underground stem and the fronds grow off of the rhizome. This means when you encounter a large mass of ferns in the woods you are basically encountering an individual plant. If you were to pick the fronds from this large mass, it would be like picking the leaves off of a single tree. The sori are another particular structure found on ferns. It is the masses or clusters of sporangia that are located on the backside of the fronds. To the layperson, these sori appear to be bugs. They are not.

Objectives

1. To become aware of the ferns and their relatives and have a basic knowledge of their life cycle.
2. To be able to identify the following structures: frond, rhizome, and sori (sorus) on a variety of different ferns.
3. To be able to identify, from a prepared slide, the archegonia and antheridia of a fern.
4. To recognize the following fern relatives: whisk fern, horsetail and club moss.

Materials

1. Several preserved and living ferns with fronds, rhizomes, and sori.
2. Preserved specimens of the whisk fern, horsetail and club moss.
3. Prepared slides of the following: fern archegonia, fern antheridia, cross-section of a frond with sori and fern rhizome.
4. Lens paper

Procedure A: Ferns (Pteridophyta)

1. Examine the preserved and living fern fronds. Compare and notice the great diversity in the different fern fronds. Look for the brown structures on the underside of the fronds. In some cases they are brown dots and in other cases they are brown bands. These are the sori. Sometimes the sori are scattered all over the backside of the frond and other times they are in a very distinct pattern, depending on the species. Sketch a frond with sori on their backside.

2. Examine a prepared slide of a cross section of a fern leaflet with sori (sporangia). The sori are clusters of sporangia. Each sporangium contains hundreds of spores. On your slide the sori are located near the margin of the leaflet. Occasionally people mistake the sori for bugs. However, they are reproductive structures. Sketch the cross section of the leaflet including the sori.

3. Examine a prepared slide of a fern gametophyte (Prothallia) with antheridia. Antheridia are the male structures that produce sperm cells. They are scattered across the gametophyte and appear as small little balls. The sperm of the fern are similar to the sperm of the moss in that they are mobile. When the sperm are mature and it rains enough to produce a film of water on the gametophyte the sperm are released and they swim to an egg to fertilize it. Compare what you see with the fern life cycle. Figure 5.1.

4. Examine a prepared slide of a fern gametophyte (Prothallia) with archegonia. Archegonia are the female portion and each contains a single egg. The archegonia are clustered near the notch of the gametophyte. When the sperm reaches the egg the egg is fertilized. The resulting zygote from that fertilization will develop into a new fern. Compare what you see with the fern life cycle. Figure 5.1.

5. Examine a prepared slide of a fern rhizome. The rhizome is an underground stem, which the roots and the shoots grow from. The rhizome is very large. When you

use low power, the rhizome will completely fill the entire field of view. As you scan around the rhizome you will notice some thick cells that are stained very dark. These cells were lacking in the mosses and are the vascular tissue. Sketch a portion of the rhizome.

Procedure B: Whisk Fern (Psilotophyta)

Examine the living specimen of the whisk fern (*Psiloum sp.*). This particular species is native to parts of Asia. It is one of our primitive land plants. Examine this plant closely. You will notice it has no leaves and it has no roots.

    a.   Where does this plant do photosynthesis? _______________________

    b.   How does this plant absorb water? (Answer is in the textbook)__________
____

Procedure C: Horsetail (Equisetophyta)

Examine the preserved specimens of the horsetail (*Equisetum sp.*). These particular species are common throughout the majority of North America. The horsetails are generally found near bodies of water or in moist forests. These plants also produce spores. The spores are usually found on the terminal end of the stems in a structure called strobilus.

Procedure D: Club Moss (Lycopodophyta)

Examine the preserved specimens of the club moss (*Lycopodium sp.*). The club mosses are found in forests stemming from the boreal forest to tropical forests. Many of the club mosses bear their spores on stalked strobili that resemble clubs.

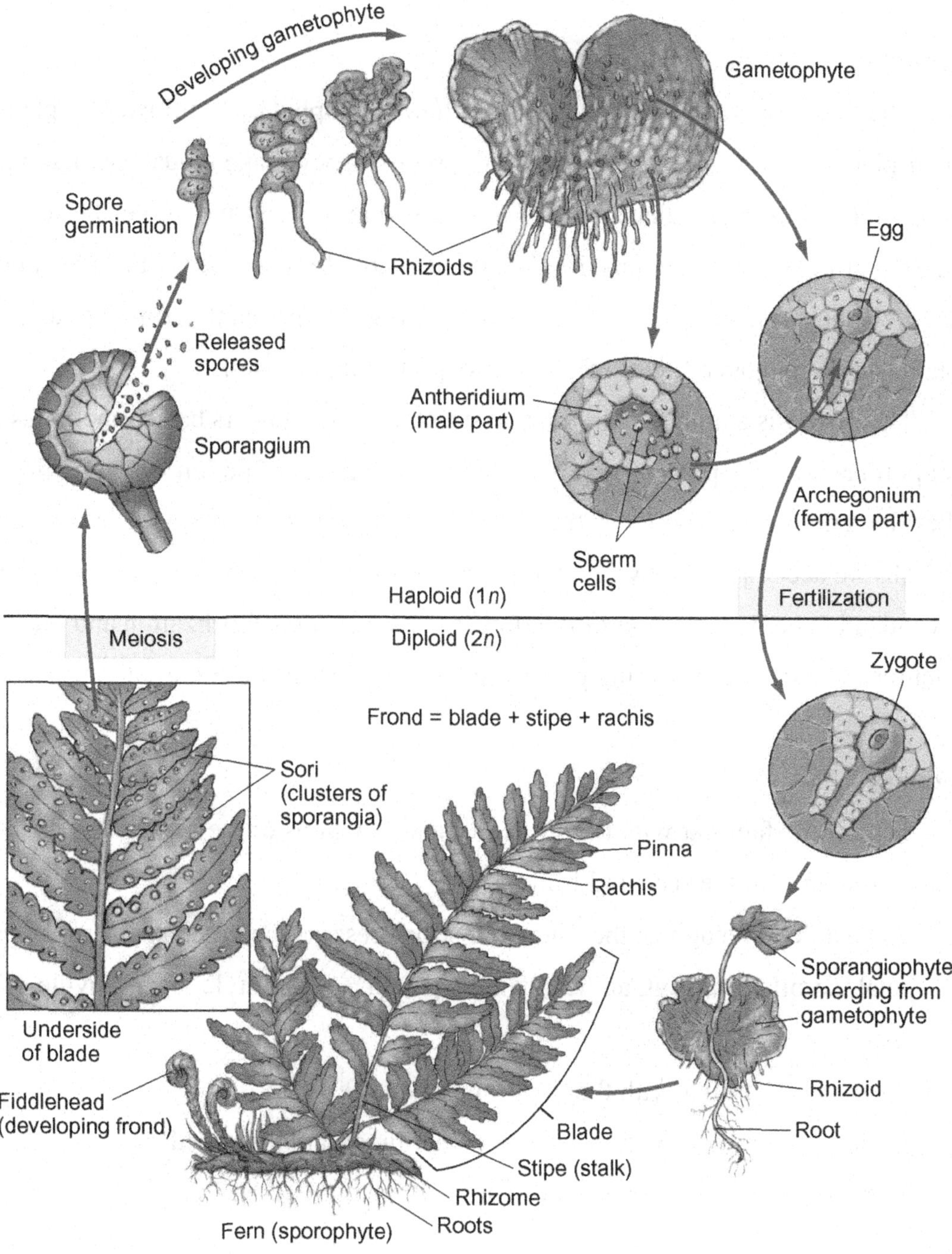

Figure 5.1 Fern Life Cycle

Laboratory 6

Flowering Plant Reproduction: Flowers, Fruits and Seeds

Introduction

The flowering plants belong to the phylum Anthophyta. The flowering plants are vascular plants; however they do not produce spores. They make seeds. By making seed the flowering plants ensure that some of their offspring will survive. A seed is a specialized structure that contains the embryonic plant. In most seeds, the largest portion is the cotyledon, which is stored energy that is used when the plant germinates and before the plant can do photosynthesis sufficient enough to support itself.

The flower is a unique reproductive structure. The flowers basic function is to attract pollinators. The process of pollination is the transfer of pollen from one flower to another flower. Once the pollen is transferred, fertilization can take place. Pollination can usually be accomplished by several means. Normally it is done by: insects, birds, bats, wind, and people. Because of the flower, which ensures fertilization and the production of seeds, the flowering plants are the most successful group of plants.

Objectives

1. To become familiar with the structures and functions of the flower and its relationship to the seed and fruit.
2. To be able to recognize the following structures: pedicel, receptacle, sepal, calyx, petal, corolla, filament, anther, stamen, pollen, stigma, style, ovary, ovule, and pistil.
3. To become familiar with the flowering plant life cycle.
4. To become familiar with the structure and function of the seed.
5. To become familiar with the structure and function of the fruit.

Materials

1. Flower Model
2. Live specimens of flowers (Gladiolus)
3. Lens paper
4. Microscope slides
5. Cover slips
6. Fresh apple
7. Lima bean seeds
8. Prepared slides: immature lily anther showing the microsporocyte, pollen tetrad, megasporocyte, mature ovule

Procedure A: Movie *The Private Lives of Plants: The Birds and the Bees* narrated by David Attenborough

This movie will introduce many aspects of flower pollination that we are not capable of seeing in our lab. This movie will cover in detail how many different plants are pollinated. Pay close attention to the different way plants can be pollinated and list as many different ways that plants are pollinated that you can detect from the movie.

1. _______________________	7. _______________________
2. _______________________	8. _______________________
3. _______________________	9. _______________________
4. _______________________	10. _______________________
5. _______________________	11. _______________________
6. _______________________	12. _______________________

Procedure B: Flowers

1. Obtain a flower and compare it to figure 6.1. Compare it to the flower model. Locate the pedicel. The pedicel is the stem of the flower. The distal end of the pedicel is slightly enlarged this is called the receptacle. The receptacle is where all of the flower parts are attached. In some flowers there is a slight modification where the receptacle is enclosed in the base of the ovary. The first flora appendages are the sepals. They are modified leaves and usually resemble leaves. However, in some instances they resemble petals. All of the sepals collectively are called the calyx. The petals are the second flora appendages. All of the petals collectively are called the corolla. The function of the corolla is to attract a pollinator. The male portion of the flower is called the stamen. The stamen is composed of two parts, the anther and the filament. The filament is a long stem like structure that holds the anther up and out of the flower. The anther is a sac-like structure on the end of the filament. It produces pollen. The final structure of the flower is the female portion of the flower, the pistil. The pistil consists of three parts. They are the stigma, style and the ovary. At the tip of the pistil is the stigma. The stigma is sticky and this is where pollen is deposited during pollination. The style is a tube-like structure between the stigma and the ovary. The large swelling at the bottom of the pistil is the ovary. The ovary is where the seeds will develop. The ovary will eventually, after pollination, mature into fruit.

    To dissect the flower you must carefully remove each of the flora parts mentioned above as you encounter them, starting with the calyx. Make separate piles of each flora part without destroying any of them. Once you have removed the calyx, corolla and stamens you should have the only the pistil left. The stamens will be used later in lab. Carefully cut into the ovary of the pistil and you will see small, light colored structures called ovules. These ovules are immature seeds. Each ovule contains one egg, which will be fertilized after pollination and then will develop into seeds.

2. Examine a prepared slide of an immature lily anther going through meiosis. In this slide you will see a cross section of an anther. It has four chambers. Inside each of these chambers is where pollen will be produced. This slide shows the beginning of this

process. The small structures on the inside of the chambers are called microsporocytes.
Figure 6.2.

3. Examine a prepared slide showing pollen tetrads. A pollen tetrad is found in the pollen
   chambers of the anther. The tetrads are the results of meiosis of the microsporocytes.
   Each quarter of the tetrad is called a microspore. Each microspore will mature into a
   pollen grain. Figure 6.2.

4. Make a wet mount of pollen. To do this you must use an anther from your dissected
   flower. Sketch and note the pollen size, shape, color and abundance from one another.

5. Examine a prepared slide of an immature lily ovary going through meiosis. In this slide
   you will see a cross section of an ovary. Each cross section of the ovary has the potential
   to contain six immature ovules. Each of these immature ovules contains one
   megasporocyte. Figure 6.2. These megasporocytes will go through meiosis to produce a
   mature ovule. When this process is completed the mature ovule will have eight nuclei.
   Figure 6.2.

6. Examine a prepared slide of a lily ovary with mature ovules. At this stage there should
   be eight distinct nuclei within the ovule. If there are only four the process is not
   complete, there must be one more division. Only one of these nuclei is the egg. It is
   usually larger than the others and situated to one end of the ovule. The other seven nuclei
   are used as an energy source to produce the seed or as an energy source used immediately
   after the seed germinates. Figure 6.2.

Procedure C: Seeds

Examine the lima bean seeds and compare it to the diagram. Figure 6.3. The seed develops from the ovule within the ovary. Seed are very important for the survival and continuance of a species. The particular structures you are responsible are: the seed coat, embryo, cotyledon, and the hilum. The seed coat is the outer protective covering. This is what allows some seeds to survive for many years. The seed coat in some plants actually inhibits germination. In this case the seed coat has to be broken down by environmental factors (wind, water, etc.) before the seed coat weakens and allows the new plant to grow. The embryo is actually a small portion of the entire seed. This is where the entire new plant begins to grow. The cotyledon, sometimes called endosperm, is basically stored energy. This is energy that is used when the seed germinates and before the plant can do photosynthesis efficiently enough to support the new plant. Finally, the hilum is where the seed was attached in the fruit. In the lima beans case the fruit was a pod.

Procedure D: Fruit

Examine the two apples, one has been cut longitudinally and the other has been cut in cross section. Figure 6.4. The apple is an unusual fruit because it has an extra layer called the floral tube. The floral tube is the portion of the apple that we eat. The core, the part we do not usually eat, is the ovary. Inside the ovary is where the seeds are found. Find the floral tube, the ovary, the ovary wall and the seeds. In other fruits the floral tube is missing as in oranges and peaches. The entire fruit is the ovary with the seeds.

    a. Is a seedless grape still a fruit? Why or why not?

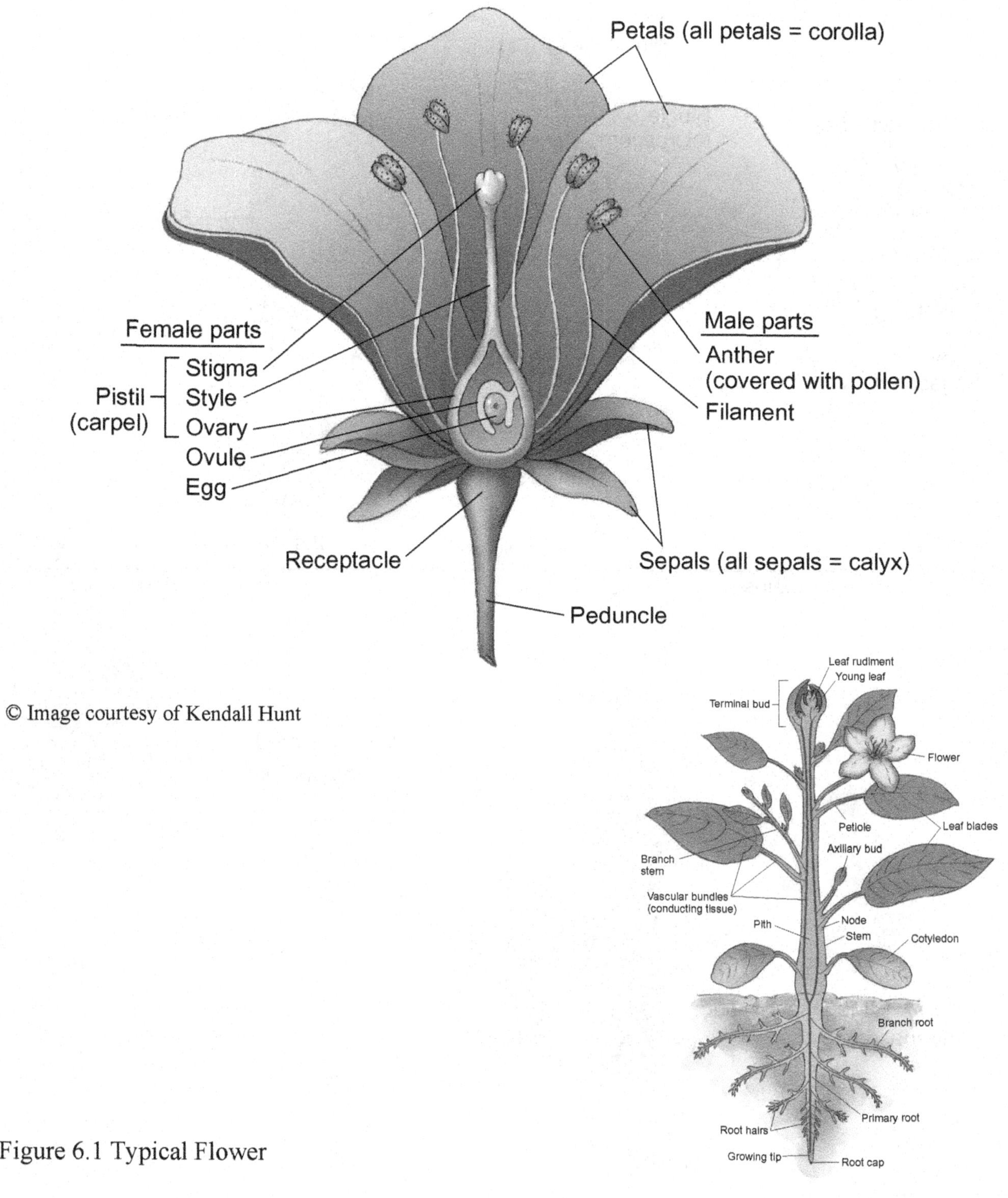

© Image courtesy of Kendall Hunt

Figure 6.1 Typical Flower

© Image courtesy of Kendall Hunt

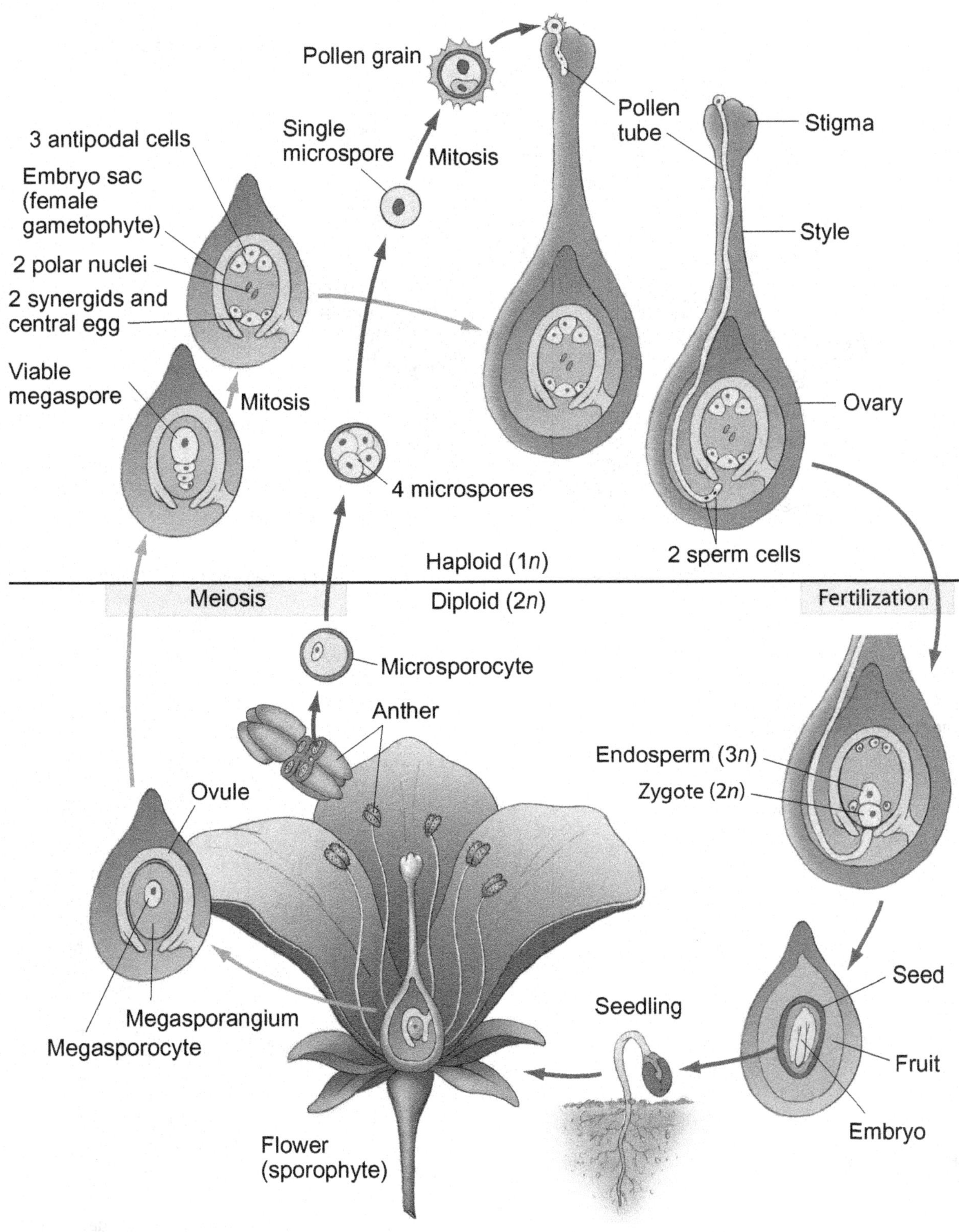

© Image courtesy of Kendall Hunt

Figure 6.2 Flowering Plant Life Cycle

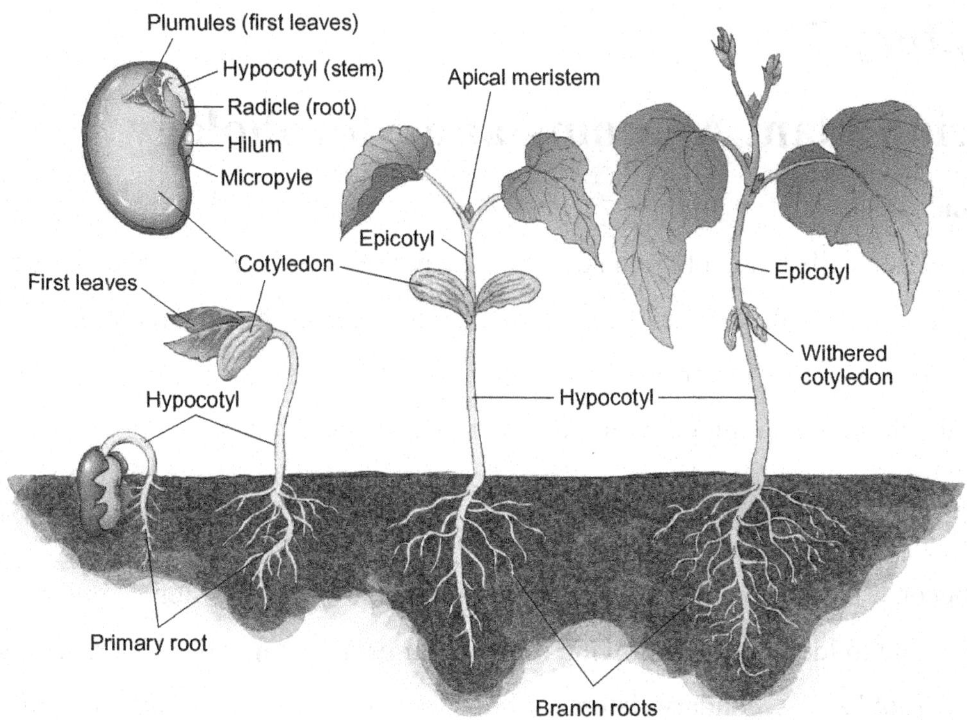

Figure 6.3 Lima Bean Seed and Seedling

© Image courtesy of Kendall Hunt

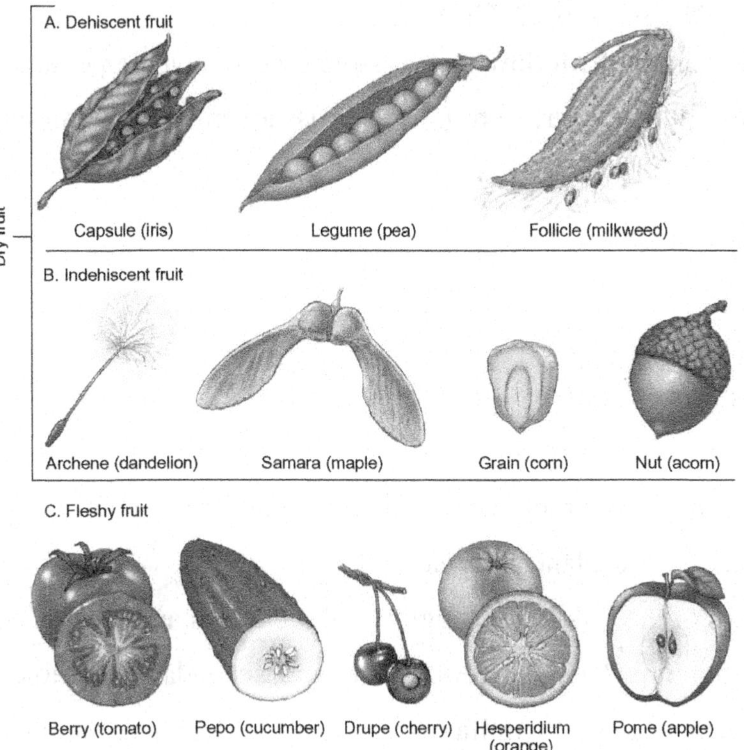

Figure 6.4 Typical Fleshy Fruit

© Image courtesy of Kendall Hunt

Laboratory 7

Flowering Plant Anatomy and Morphology

Introduction

Specialized cells in the phylum Anthophyta are arranged like other specialized cells in other phyla, into tissue. These tissues are arranged into organs. We already looked at the reproductive organs of the flowering plants. They were the flower, fruit and seed. In this lab we will look at the three remaining plant organs, the roots, stems, and leaves.

Objectives

1. To become familiar with the structure and function of the root, stems, and leaves.
2. To be able to identify the following root structures: root cap, apical meristem, vascular tissue, root hairs, secondary (lateral or branch) root, cortex, region of elongation, and the region of maturation.
3. To be able to identify the following stem structures: apical meristem, leaf primordia, vascular tissue and the axillary buds.
4. To be able to identify the following leaf structures: cuticle, upper and lower epidermis, palisade layer, spongy layer, veins (vascular tissue), air pockets, stomata, guard cell and stoma.

Materials

1. Lens paper
2. Microscope slides
3. Cover slips
4. Models of the following: root, dicot stem, monocot stem, woody stem and a leaf.
5. Living Wandering Jew plant (*Zabrina sp.*)
6. Prepared slides of the following: longitudinal section of an onion (*Alium sp.*), cross section of Buttercup (*Rannuculus sp.*), root hairs, secondary root, longitudinal section of *Coleus sp.*, cross section of Alfalfa (*Medicago sp.*), cross section of corn (*Zea mays*), cross section of basswood (*Tilia sp.*), cross section of lilac (*Syringia sp.*) and a cross section of pine needles.

Procedure A: Roots

The roots are specialized underground organs. There are two basic types of roots, fibrous and tap. The fibrous roots are a mass of entangled roots found in the grasses. Tap roots are central, large and fleshy, as in a carrot. The first function of the root is absorption. The roots are responsible for absorbing water and minerals. The second function of the root is anchorage. The roots are responsible for holding the plant in place. The third function of the root is storage. The excess nutrients that are produced from photosynthesis and other substances such as water are stored in the roots.

1. Examine a prepared slide of the onion (*Alium sp.*) root tip. Compare it with the diagram, figure 7.1, and the model. Locate the root cap. This region protects the root from abrasion as it is pushed through the soil. Next, locate the apical meristem. This is the tissue adjacent to the root cap. Its function is mitosis (cell division). Adjacent to the apical meristem is the region of elongation. In this region cells elongate and are becoming specialized to perform a specific function. The adjoining section of the region of elongation is the region of maturation. In the region of maturation these cells are fully grown and are specialized to perform specific tasks. In this region you will find the root hairs. The purpose of the root hairs is to increase the surface area of the root in order to absorb more water and minerals. In the center of the root you will find vascular tissue. The cells closest to the apical meristem are not functioning. The vascular tissue cells in the region of maturation are functioning.

2. Examine the prepared slide of a buttercup (*Rannunculus sp.*) root cross section. Compare it to the diagram, figure 7.2, and the model. In the center of this cross section you find the vascular tissue. You can see both types of vascular tissue in this slide. The xylem is the cells that are stained dark in the shape of a triangle or x at the center of the vascular tissue. The phloem is the cells that are found in between the xylem that radiates out to the edge of the vascular tissue. Between the vascular tissue and the outer edge, the epidermis, are cells called the cortex. These are specialized cells for storage.

3. Examine a prepared slide of root hairs. Root hairs are an extension of the epidermal
    cells. Their function is to increase the surface area, which allows for greater absorption
    of water and minerals. Sketch the root with root hairs.

4. Examine a prepared slide of a cross section of branching roots. These branched roots are
    also called lateral roots and secondary roots; basically these are new roots. The branched
    root arises from the vascular tissue. The branched root grows out through the cortex and
    epidermis. Sketch a cross section of a stem showing a new-branched root.

Procedure B: Stems

The basic function of all stems is to enable plants to compete for light in three dimensions.
Essentially it is an attempt to outcompete other plants for sunlight by being taller and out of other
plants shade. By doing this, photosynthesis is more efficient. There are two basic variations of
stems we will look at, herbaceous and woody.

1. Examine a prepared slide *Coleus sp.* stem tip. Compare it with the diagram, hand out.
    The apical meristem (cell division) is found at the top of the stem tip. Note the absence
    of a cap. At both sides of the apical meristem are protrusions. These are leaf primordia

(developing leaves). In between the developing leaves, are the axillary buds. These are buds that will develop into flowers or new branches.

2. Examine a prepared slide cross section of alfalfa (*Medicago sp*) stem and compare it to the diagram, figure 7.2, and the model. There are two types herbaceous stems. The alfalfa represents the first type called a dicot stem. A dicot stem is distinct from the other by its position of vascular tissue. The vascular tissue is arranged in bundles <u>just</u> <u>under</u> <u>the</u> <u>epidermis</u>.

3. Examine a prepared slide of corn (*Zea mays*) stem cross section and compare it to the diagram, figure 7.3, and the model. The corn represents the second type of herbaceous stem. This is a monocot stem. A monocot stem is distinct from the dicot stem by having vascular tissue arranged in bundles and <u>scattered</u> <u>throughout</u> the cross section.

4. Examine a prepared slide of basswood (*Tillia sp.*) stem cross section. Basswood has a woody stem, compare it to the model. It has distinct rings of dead xylem (vascular tissue) with a small portion of living xylem under the bark. The phloem (vascular tissue) makes up the inner bark. The outer bark is cork, not living. In temperate regions each ring represents a single growing season.   In tropical areas each ring may or may not represent a single growing season. They usually represent wet and dry seasons. Sketch the *Tillia* cross-section and label.

Procedure C: Leaves

Leaves are broad surface structures that function in collecting light in order to do photosynthesis. Another function is gas exchange. Carbon dioxide, CO_2, is needed for photosynthesis and is gained from the atmosphere. Oxygen, O_2, is a waste product of photosynthesis and needs to be removed.

1. Examine a prepared slide of a lilac (*Syringia sp.*) leaf cross section. Compare it to the diagram, figure 7.4, and model. Locate the following structures. The cuticle is a waxy substance that is found on the upper and lower surface of the leaf. Its function is to prevent desiccation, protection from drying out. Towards the inside of the cuticle is the epidermis both on the upper and lower surfaces. The palisade layer is the layer of cells directly under the epidermis. These cells contain numerous chloroplasts and do most of the photosynthesis in the leaf. Under the palisade layer is the spongy layer. This layer contains many air pockets. The air pockets function is gas storage. Periodically, in between the palisade layer and the spongy layer there are bundles of vascular tissue. These are called veins. The last structure is the stomata. The stomata are found on both the lower and upper epidermises. The stomata function is gas exchange regulation. The stomata have two parts. The two parts are the guard cells, which regulate the gases and the stoma, which is the opening between the guard cells through which the gases pass.

2. Using the technique demonstrated by the instructor, remove the lower epidermis from one of the leaves, which has been placed in distilled water, and prepare a wet mount. Notice the thick inner walls of the guard cells and the thin outer walls. You may also notice that the other epidermal cells do not have chloroplasts. Sketch the guard cells and stoma. Note the size of the stoma.

3. Examine a prepared slide of a pine needle cross section. Note the differences between the pine needle and the lilac leaf. Sketch, noting the position of the epidermis and the vascular tissue.

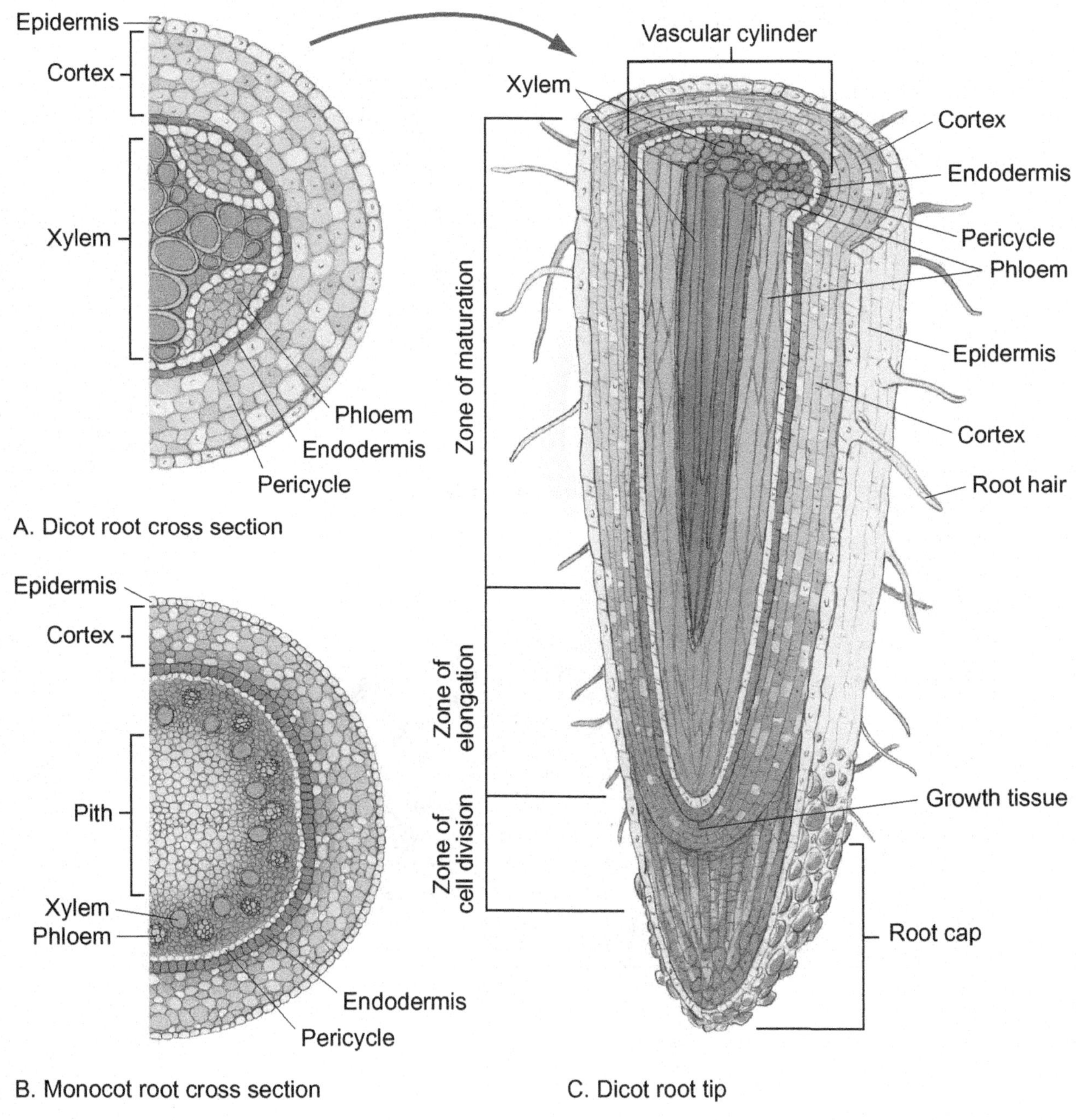

© Image courtesy of Kendall Hunt

Figure 7.1 Typical Root Structures

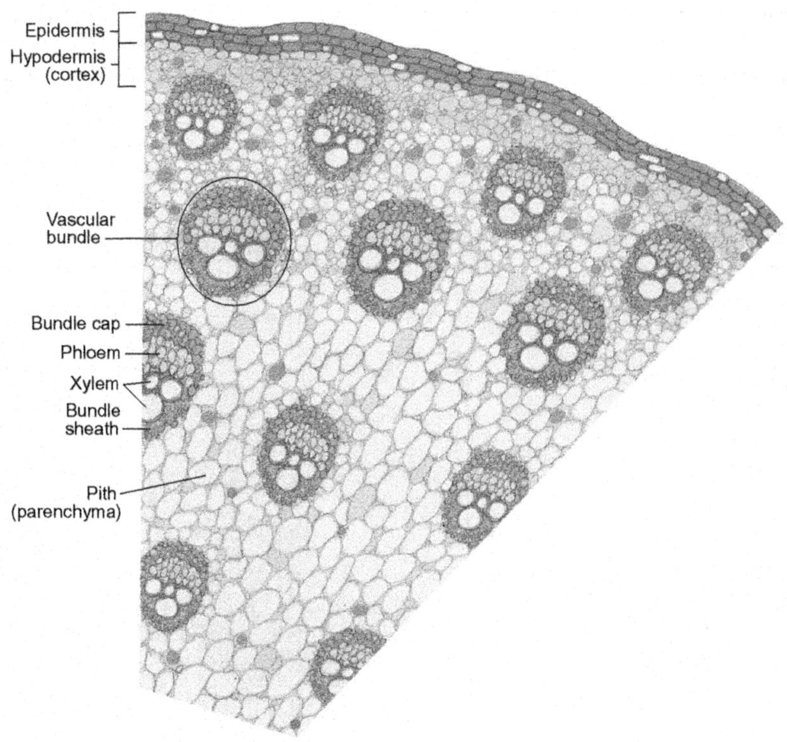

© Image courtesy of Kendall Hunt

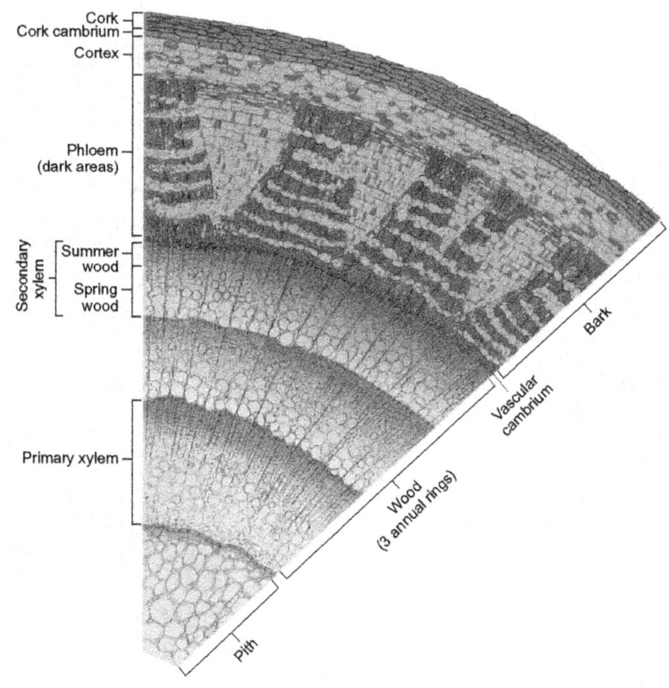

Figure 7.2 Dicot Stem Cross-section

© Image courtesy of Kendall Hunt

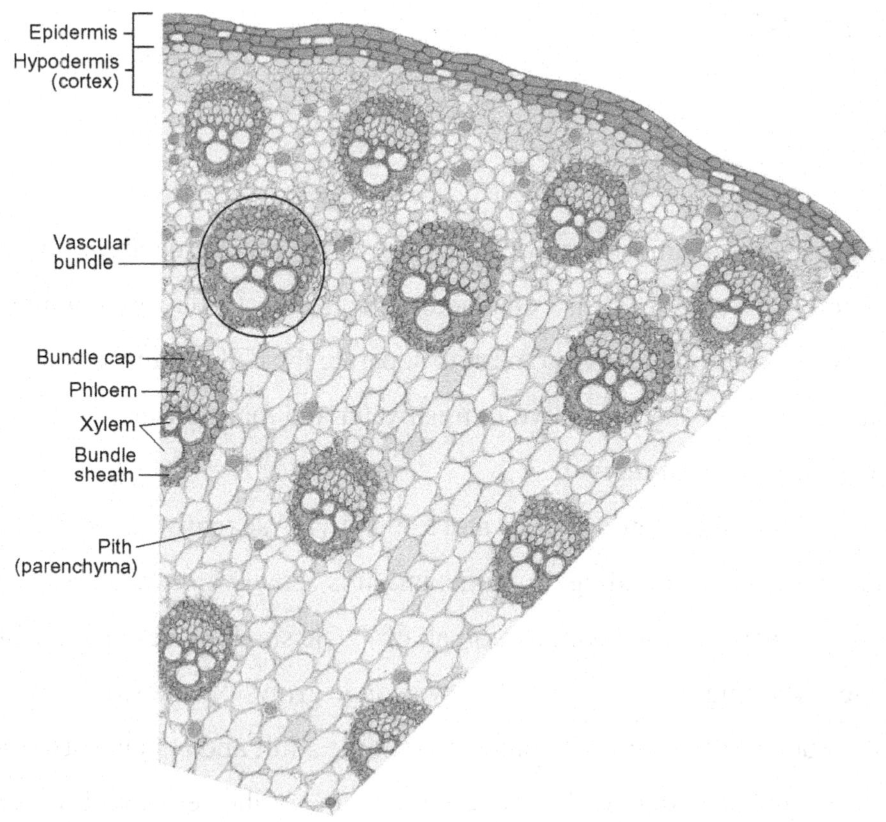

© Image courtesy of Kendall Hunt

Figure 7.3 Monocot Stem Cross-section

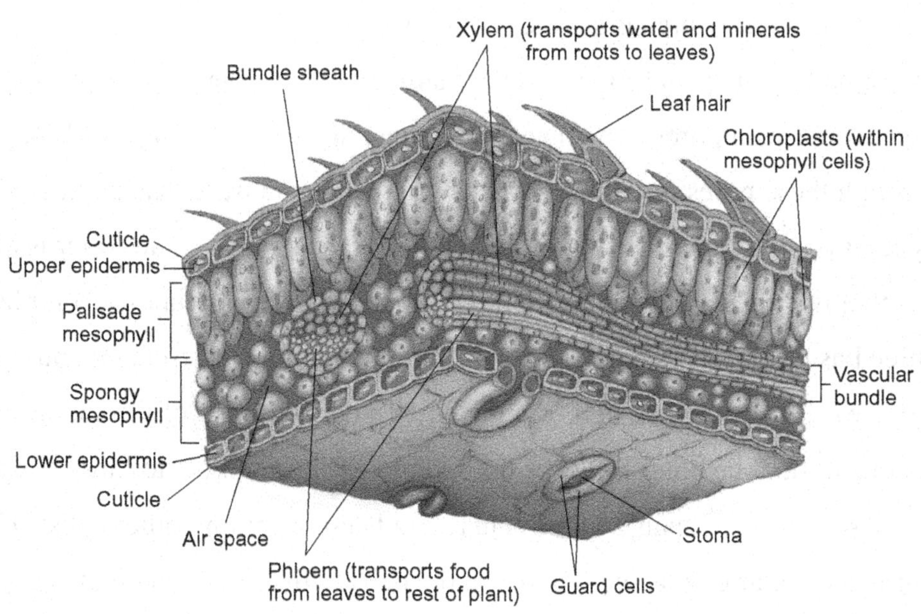

© Image courtesy of Kendall Hunt

Figure 7.4 Leaf Cross-section

Laboratory 8

Porifera

Introduction

Sponges are one of the most primitive animals. They are multicellular, however in many ways they do resemble colonies of cells. One oddity that suggests they are colonies is that some sponges can be squeezed through a nylon mesh to separate the cells and then they can reassemble back into a sponge again. Sponges can live in freshwater and marine water; however, most of them are marine, living on coral reefs.

Sponges can reproduce both sexually and asexually. Asexually, the sponge can reproduce by fragmentation or budding. Fragmentation is usually caused by wave action. The waves cause a small portion of the sponge to break off. The small pieces are capable of growing into a new sponge. Budding is where a small new sponge grows off the side of the original adult sponge. Once the bud is fully formed it detaches from the adult and attaches itself to the bottom and matures. The sexual reproduction is a function of some of the cells that line central cavity. Some of these cells produce sperm and others produce eggs. Both of these gametes are released into the water where fertilization takes place. The resulting zygote will mature into a larva that swims away from the adult, which is sessile (not moving).

The basic body plan of the sponge is very simple. As the phylum name suggests, these organisms are covered with pores, which are small openings for water to pass through. The water goes through these pores into the central cavity. Once the water has circulated through the central cavity it leaves through the osculum on the top of the sponge. The water is circulated by the collar cells that line the central cavity. The collar cells have flagella to accomplish this. The sponge has three basic cell layers: epidermal cell layer, ameobocyte cell layer and the collar cell layer (Choanocyte). The epidermal cells are the outer most cell layer of the sponges. These are flattened cells that define the outer limits of the sponge. The ameobocytes are cells that resemble *Amoeba sp.* These cells are capable of moving around and becoming other cells. The inner most cell layer is the collar cell layer. These are the cells with the flagella that cause the water to circulate. They also collect the food that floats into the sponge with the circulating water. Another function of a few of the collar cells is to produce the gametes for sexual reproduction.

The spicules are the last structures we will be concerned with. They are glass-like spines that function in support and protection.

Objectives

1. To become aware of the structures and functions of the various parts of the sponge.
2. To be able to identify the following structures from a prepared slide: central cavity, pores, osculum, epidermal cells, amoebocytes, collar cells, and spicules.

Materials

1. Lens paper
2. Preserved specimens of living sponges
3. Prepared slides of the following: *Grantia sp.* and spicules.

Procedure A:

1. Examine the various preserved specimens of sponges, which are on display. These are actually the remains of former living sponges. The living cells of these sponges have decomposed and what is left is the sponge skeleton, which is made of spicules and spongin. The spongin is a network of fibers along with the spicules that are for support and protection in the sponge.

2. Examine a prepared slide of a *Grantia sp.* cross-section. Compare the slide with the diagram, figure 8.1. The opening in the center of the sponge on the slide is the central cavity. Most of the cells that you see that outline the central cavity are collar cells. You may also see lines of collar cells that radiate out from the central cavity to the epidermal cells. These collar cells outline the pores which water enters the sponge through. This small little sponge has no collar cells and no amoebocytes that can be seen.

3. Examine a prepared slide of sponge spicules. These are glass-like spines. If you look closely, you will see there are two types of spicules. One is like a needle with points on both ends. The other type has three arms. Each of the arms has sharp points. Sketch a few spicules.

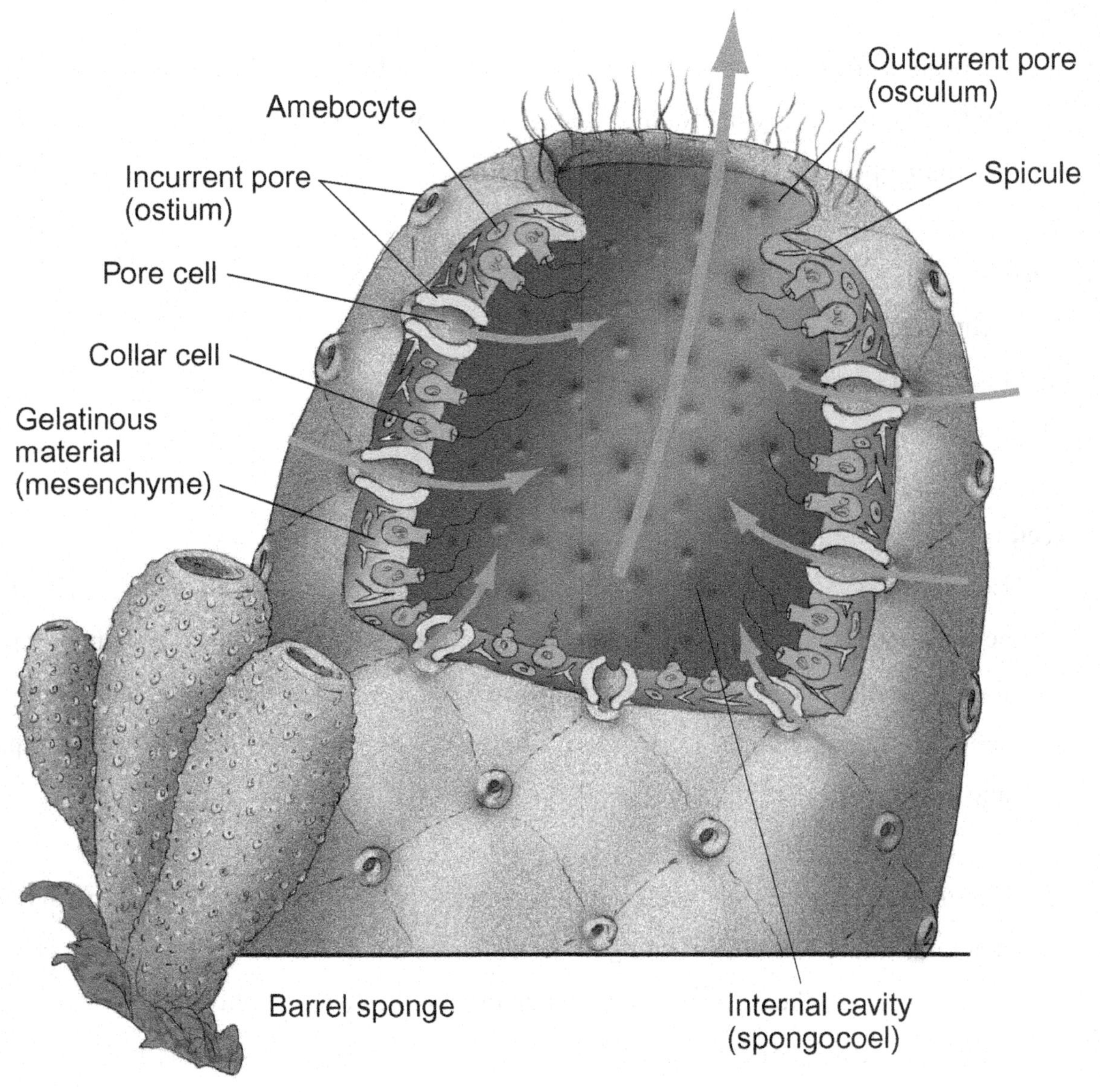

© Image courtesy of Kendall Hunt

Figure 8.1 Sponge

Laboratory 9

Cnidaria

Introduction

The Cnidarians are a diverse group of organisms. They can live in freshwater or in marine water. None of the living Cnidarians are terrestrial. The living Cnidarians have two basic body plans. One is called a polyp and the other is called a medusa. The polyp is a cylindrical bodied organism with tentacles located around the mouth on top of the organism. The medusa is a jellyfish. Its mouth is on its underside and the tentacles hang down around it. Some Cnidarians are polyps their entire lives, other Cnidarians are medusas their entire life and some Cnidarians are both a polyp and a medusa depending on when it is observed in its life cycle.

All of these organisms have several structures that are in common no matter what their body type is. First they all have tentacles with nematocysts. The tentacles are for grabbing food and putting it into its mouth. The nematocysts are stinging cells that paralyze the prey in order for the tentacle to grab it. The next structures that they have in common are the cell layers. They have three basic cell layers, they are: ectoderm, endoderm and mesoglea. The ectoderm is the outside layer for protection. The endoderm is the inside layer that lines the gastrovascular cavity. This cell layer helps in digestion. The mesoglea is contractile tissue. It is for movement. The last structure that the Cnidarians have in common is the gastrovascular cavity. The gastrovascular cavity is a cavity that extends throughout the organisms body, even out through the tentacles. The function of this cavity is digestion and circulation.

Objectives

1. To become aware of the structure and function of a few examples of the phylum Cnidaria.
2. To be able to identify the different organisms that belongs to the Cnidarian phylum.
3. To be able to identify the following structures: tentacles, nematocysts, ectoderm, endoderm, mesoglea and the gastrovascular cavity.

Materials

1. Lens paper

2. Microspore slides

3. Cover slips

4. Small watch glasses

5. Plastic droppers

6. Dissecting scopes

7. Toothpicks

8. Living specimens of *Hydra sp.* and *Daphnia sp.*

9. Preserved specimens of: *Obelia sp.*, jellyfish, Portuguese man-of-war, sea anemone, coral and Petoskey stone.

10. Prepared microscope slides of: *Hydra sp.* (Cross section and Longitudinal section), *Obelia sp.*, nematocysts, *Obelia* medusa and *Obelia* planula.

Procedure A: Hydra

The *Hydra sp.* is a freshwater form of Cnidaria, which is found in many streams, ponds, and lakes.

1. Obtain a living *Hydra* and place it in a small watch glass. Be careful, because the *Hydra* gets stuck in the plastic dropper very easily. Observe the *Hydra* under the dissecting microscope. The *Hydra* moves slowly and it may attach itself to the side of the glass or to the surface of the water so it appears upside down. This is a small polyp with a few tentacles on one end. Compare the *Hydra* with figure 9.1.

2. While watching the *Hydra* under the dissection microscope, gently touch the *Hydra* with a toothpick. Do this gently at first and then a little more forcefully. Be careful not damage the *Hydra*.

    a. What does the *Hydra* do when it is stimulated with the toothpick?

3. Place some *Daphnia sp.* into the watch glass with the *Hydra*. The *Daphnia* are a food source for the *Hydra*. You may notice that some of the *Daphnia* are actually larger than the *Hydra*. The size of the *Daphnia* does not matter. If the *Hydra* can sting it with the

nematocysts on its tentacles, it will try and put it into its mouth. Observe what the *Hydra* does with the *Daphnia*.

4. Prepare a wet mount of the living *Hydra*. Place the *Hydra* on the slide and carefully place the cover slip over it. Do not push down on the cover slip because you will smash the *Hydra*. Notice the way the Hydra moves. You may notice some thread-like structures extending from the tentacles. These are nematocysts that have been discharged.

5. Examine a prepared slide of the longitudinal section and the cross-section of the *Hydra*. Find the gastrovascular cavity, epidermis, gastrodermis and mesoglea.

6. Examine a prepared slide of nematocysts. These slides are smears with numerous nematocysts. The nematocysts are very small and you must use high-power to see them. Some of these cells will still have the stinger coiled, ready to spring out and inject its poison into a prey. Other nematocysts will have their stingers stretched out and discharged. Sketch a few nematocysts.

7. Examine a prepared slide of a *Hydra* reproducing asexually by budding. This is where the *Hydra* grows a new little *Hydra* off of its side.

Procedure B: Obelia

The *Obelia sp.* is a marine Cnidarian. It is usually attached to rocks; shells and other things that do not float around very easily.

1. Examine the preserved specimens of the *Obelia*. The *Obelia* is an organism that is actually a colony of polyps. Sketch the general pattern of the branching.

2. Examine a prepared slide of an *Obelia* colony and compare it to the diagram, figure 9.2. There are two types of polyps. The first one has tentacles; it is called the feeding polyp. The second polyp is the reproductive polyp and it has no tentacles.

    The feeding polyps have a gastrovascular cavity. It is continuous through all of the polyps in the colony. The reproductive polyps do not feed; they do not have any tentacles to catch food. They do however receive nutrients because their gastrovascular cavity is connected to the other polyp's gastrovascular cavity. The reproductive polyp reproduces asexually by budding off small little medusas (jellyfish). These medusas are either male or female and can reproduce sexually. The female releasing eggs and the male releasing sperm into the water accomplish this, if the sperm and eggs are released in close proximity, fertilization will take place. When the sperm and the egg fuse together they become a zygote. The zygote matures into the embryo and the embryo matures into a larval stage called the planula. The planula swims around seeking out a suitable habitat that has sufficient food floating in the water. When a suitable habitat is found the planula will attach to the bottom and grow into the mature colony.

3. Examine a prepared slide of an *Obelia* medusa. The function of the medusa is dispersal from the mature colony and sexual reproduction. Sketch the medusa.

4. Examine the prepared slide of the planula. The planula is the free-swimming larva that will eventually develop into the mature colony. Sketch the planula.

Procedure C: Jellyfish

Observe the preserved specimens of the jellyfish. Jellyfish are medusas. Their polyp stage is quite small and insignificant. Jellyfish are also a food source for larger animals. Remember, if you find a jellyfish on an ocean beach; do not touch it. The dead jellyfish can still produce a very painful sting.

Procedure D: Portuguese Man-of-war

Examine the preserved specimen of the Portuguese man-of-war. This is not a jellyfish. It is a colony of polyps with a large gas filled bladder. The air bladder is what allows the Portuguese man-of-war to float. The long tentacles that hang down from the air bladder have numerous nematocysts and catch a large number of prey.

Procedure E: Sea Anemone

Observe the preserved specimen of a sea anemone. This is a large polyp with very large tentacles and a lot of nematocysts. The sea anemone is so large that it has internal structural support. In addition, the sea anemone forces water into its body to create a slight pressure to keep it from collapsing.

Procedure F: Coral

Examine the preserved specimens of coral. The secretions from the coral polyps create coral. These secretions harden to from a rock-like substance that the polyp uses for protection. The polyps emerge from these protective hideouts with their tentacles out to collect food from the water.

Procedure G: Petoskey Stone

Examine the Petoskey stone. It is the fossilized remains of coral and Michigan State Stone. It is evidence that Michigan was once under a shallow topical ocean. If you look closely at a wet Petoskey stone, you can see where the individual polyps lived. Sketch the Petoskey stone.

Figure 9.1 *Hydra sp.*

Figure 9.2 *Obelia sp.* Life Cycle

Laboratory 10

Rotifera

Introduction

Rotifers are small microscopic multicellular organisms. They are common in freshwater and are sometimes found in marine water. They can also be parasites. When examining freshwater under the microscope from around our local area, it is not uncommon to find rotifers. It is easy to mistaken them for protozoa. However, they are not. Remember they are multicellular and not single-celled like the protozoa.

Rotifers are small little carnivores. They will eat almost anything that is small enough. When observing the living specimens, you will notice on one end, a rotating device. This structure really is not rotating, it just appears that way. The structure is call a corona and is cover with cilia. The function is to pull food into the organism to be ground up and digested.

Objectives

1. To be able to distinguish the difference between the rotifers and their food source, i.e. protozoa
2. To be able to recognize the rotifer as a living and preserved specimens.

Materials

1. Lens paper
2. Microscope slides
3. Cover slips
4. Watch glasses
5. Plastic droppers
6. Prepared slides of rotifers
7. Living specimens of rotifers
8. Living *Daphnia sp.* to feed the rotifers

Procedure A: Rotifers

1. Examine a preserved slide of rotifers. Compare the rotifers on the slide with the diagram of the rotifer, hand out.

2. Examine the living specimens of rotifers under the dissecting microscope. Place a few rotifers in a watch glass and observe their movement.

3. Place some *Daphnia* into the watch glass and observe what the rotifer does. The *Daphnia* are food for the rotifer. The rotifers will eventually eat the *Daphnia*. Watch the cilia to see how they take the food in. Sometimes you will be able to see the *Daphnia* on the inside of the rotifer.

Laboratory 11
Mollusca

Introduction

The phylum Mollusca is a very large phylum with more than 100,000 different species, including snails, slugs, clams, squids, and octopus, etc. Clams are bivalve mollusks, meaning they have two valves (shells) that are hinged in a way that the two valves can open. Clams are found in both freshwater and marine habitats from the equator to both poles. They can range in size from being almost microscopic to the giant clams that can weigh nearly 200 kg (over 400 pounds). All clams are filter feeders, meaning they siphon organic materials from the water and digest it for their nutrients and all clams show no cephalization, meaning they lack a definitive head end. Many species of clams are used food by different cultures around the world. However, many of these clam species are too small to be eaten by humans and not all of them are palatable.

Objectives

1. To recognize the various groups within the phylum Mollusca.
2. To dissect a clam and be able to identify the various internal and external structures.

Materials

1. Dissecting trays
2. Dissecting tools
3. Preserved specimens of clams for dissecting
4. Standard screwdrivers

Procedure

Place the clam in the dissecting tray and locate the **anterior, posterior, dorsal, ventral,** and **lateral surfaces** (Figure 11.1). Your clam has two **valves** (shell halves). Place the clam in the dissecting tray with the left valve up. To decide which is the left valve or right valve, look at the

umbo which is the bump at the anterior end of the valve. Orient your clam so that the umbo is to the upper, left side (see diagram) and the left value will be up.

Pick up the clam and hold it so that the dorsal side is down. Carefully insert a screwdriver between the ventral edges of the two valves. Be very careful not to destroy the internal structure of the clam and not to jam the screwdriver into your hand. Next, use the screwdriver to pry open the valves so they are about one centimeter apart. Now, locate the **adductor muscles** (Figure 11.1). The adductor muscles are what control the opening and closing of the two valves; they are what hold the two valves closed. Cut the anterior adductor muscle first, as close to the inside of the left valve as possible. You do this by placing the scalp between the shell and the **mantle** and cutting the **anterior adductor muscle** away from the left shell. Repeat this procedure to remove the left valve from the **posterior adductor muscle**. Once both adductor muscles have been cut open, you can bend the left valve back so that it lays flat in the dissecting tray. Be careful that you do not tear the mantle as you manipulate the left valve out of the way. At this time, the left valve has essentially been removed. Take the left valve and run your fingers over it and feel the different textures of the inner and outer surfaces.

Locate the **mantle**, this is the tissue that covers the soft body of the clam and lines the inner surface of the two valves. Also locate the **mantle cavity** the space inside the mantle.

With the clam lying in the dissecting tray, locate the two openings at the posterior end of the clam. These are the **incurrent siphon** and **excurrent siphon**. The siphon that is more ventral is the incurrent siphon and brings water into the clam. The dorsal of the two siphons is the excurrent siphon and expels water and waste from the clam. **Siphon retractor muscles** are used to retract or pull the siphons into the shell for protection.

With your scissors, carefully cut away the part of the mantle that covers the left valve. The left **gill** will be exposed now. This is used for respiration. Next, remove the rest of the left mantle along with the siphon retractor muscle, being careful not to cut the exposed gill. Cut with a scalpel lifting the mantle as you go. By doing this, you will expose the gills more and the **foot**. Near the anterior adductor muscle of the clam and near the **mouth**, you will see two flap-like structures called the **labial palps**. Clams are filter feeders. The labial palps are used my clams to aid in feeding by collecting organic material they siphon in through the incurrent siphon. The gills are used as a filter for the organic material they consume as well as the removal of oxygen from the water.

The muscular foot is ventral to the gills. It is use to burrow into the sand and mud in order to conceal themselves from predators or to prevent from drying out during low tides.

The ventral portion of the foot needs to be removed next. Carefully cut this away with the scissors and then use the scalpel to cut the muscle at the top of the foot into right and left halves. At this time you can carefully peel the muscle layer away to view the internal organs.

The yellowish, spongy material is the **reproductive organs**. The greenish structure near the umbo is the **digestive gland** and it is surrounding the **stomach**. Extending from the stomach is the **intestine**; it is usually long and coiled. Follow the intestine to an area dorsal to the posterior adductor muscle and you will find the **heart** wrapped around the intestine. Continue following the intestine to the end and find the anus, posterior to the posterior adductor muscle.

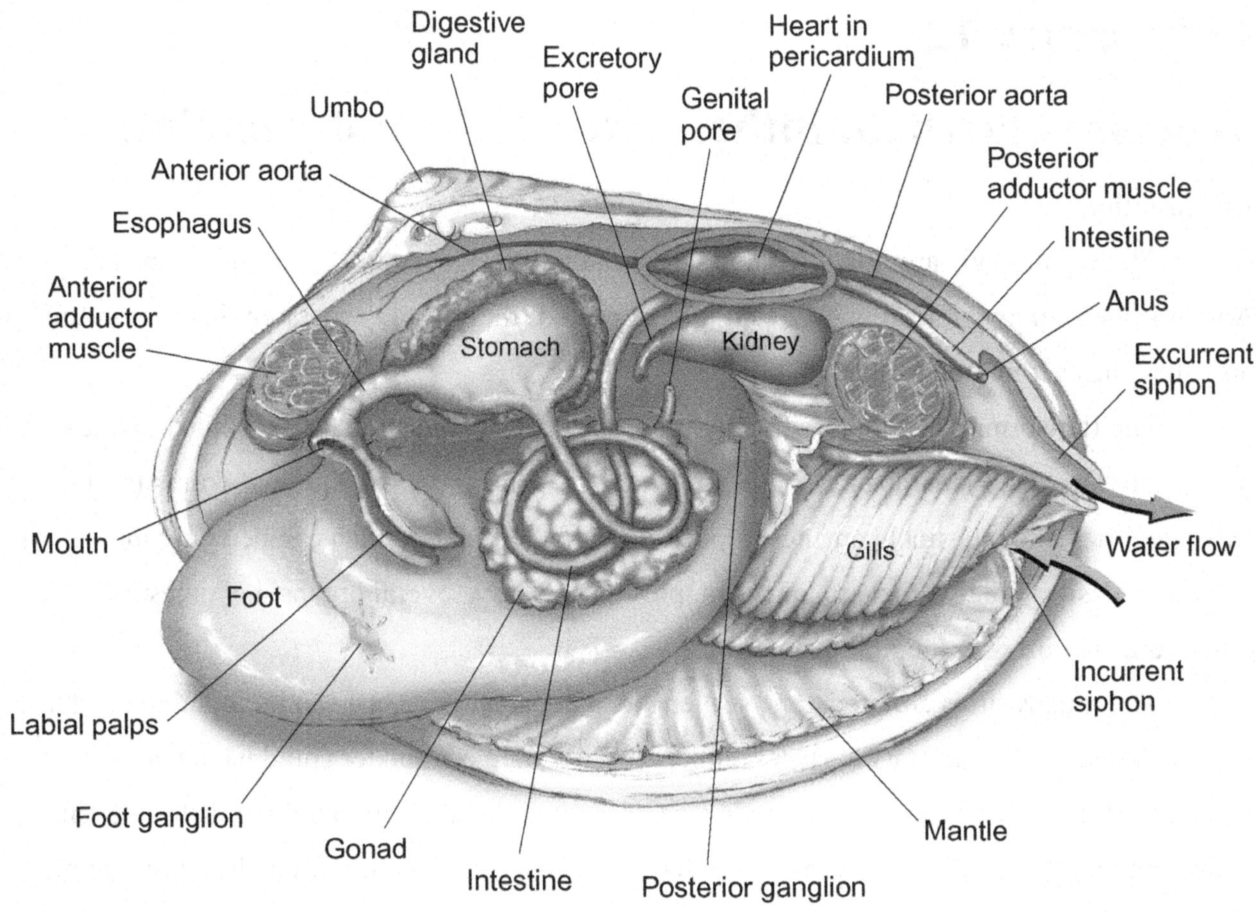

© Image courtesy of Kendall Hunt

Figure 11.1 Clam Anatomy

Laboratory 12

Worms: Platyhelminthes, Nematoda and Annelida

Introduction

The worm phyla are: Platyhelminthes the flatworms, Nematoda the roundworms and Annelida the segmented worms. We will look at each of these phyla separately in order to see how they differ from one another.

The flatworms can either be free living or they can be parasitic. We will look at one free-living flatworm, the Planaria and several parasitic flatworms. The parasitic flatworms fall into two categories: the intestinal parasites (tapeworm) and the flukes, which can be either internal or external parasites. Flatworms are fairly complex. They have several organs, i.e. digestion organs and a brain with longitudinal nerves.

The roundworms are predators, scavengers and parasites. They can be found in most any aquatic and terrestrial habitats in such numbers that a square centimeter could have tens to hundreds of individuals, depending on the size of the individual. The roundworms have even more complexity than the flatworms. They have complete organ systems, i.e. digestion. Some of these roundworms can be parasites in people. We will look at a few of these.

The segmented worms are the most complex group of the worms. They can be found in a variety of habitats. Many of them are scavengers surviving on the remains of other organisms. This is also a very large group with many individuals. We are most familiar with the earthworms, however there is some that have a long cylindrical body like an earthworm but they have pairs of appendages down their bodies (Polychaetes). We will use the earthworm as our only example of the segmented worms.

Objectives

1. To recognize the various groups of worms (Platyhelminthes, Nematoda and Annelida).
2. To be able to recognize the various structure within the different worms.
3. To understand how a person gets a parasitic worm and what precautions can be taken to avoid contracting one.
4. To dissect an earthworm and be able to identify the various structures.

Materials

1. Lens paper
2. Microscope slides
3. Cover slips
4. Watch glass
5. Plastic droppers
6. Dissecting trays
7. Dissecting tools
8. Dissecting pins
9. Living specimens of the following: planaria and vinegar eel (*Turbatrix aceti*)
10. Preserved specimens of earthworms to be dissected.
11. Model of the earthworm.
12. Prepared slides of the following: planaria, sheep liver fluke (*Fasciola hepatica*), human liver fluke (*Clonorchis sinesis*), blood fluke (*Schistosoma mansoni*), beef tapeworm (*Taenia saginata*), *Trichinella spiralis*, hookworm (*Necator americanus*) and pinworm (*Enterobius vemicularis*)

Procedure A: Platyhelminthes

1. Examine the prepared slide of a planaria. The planaria is a non-parasitic flat worm. It is commonly found in fresh water around Genesee County. Figure 12.1. Notice that the planaria has two eyespots. These are not for seeing, like our eyes, but instead they are photoreceptors for the detection of light. By detecting light they can move towards light where they will find photosynthetic organisms that they can consume. Also, you will see the pharynx. The pharynx is a small hose that extends from the midsection of the planaria that is used to draw in food. If the planaria is stained properly you will also be able to see the internal structures. Note the gastrovascular cavity. The gastrovascular cavity is a sac-like gut where food is digested.

2. Examine a living specimen of the planaria. Place a small amount of water with a planaria into a watch glass. Be careful not to get the planaria stuck in the plastic dropper. Place

the watch glass containing the planaria under a dissecting microscope and observe the planaria for a while. While observing the planaria note the way they swim.

3. Examine a prepared slide of a sheep liver fluke (*Fasciola hepatica*). All flukes are parasites in or on animals. The sheep liver fluke is primarily a parasite in domestic farm animals. However, people can also get this parasite. Figure 12.2. Even though this is a large organism, using the microscope note the mouth and the large gastrovascular cavity. Directly below the mouth you will see a large circular structure. This is a sucker that is used to attach to the host. As an adult this particular fluke lives in the bile duct of the liver. Here it lives upon blood and the tissue lining of the bile duct. The presence of this parasite causes cirrhosis of the liver.

4. Examine a prepared slide of the human liver fluke (*Clonorchis sinesis*). The human liver fluke has a life cycle that is quite similar to the sheep liver fluke. However, one of its primary hosts is people. People get this parasite by consuming raw or undercooked fish, which have been infected by the larval stage of the human liver fluke. If a person eats a raw or undercooked fish infected with the larva, the larva is absorbed into the blood and makes its way to the liver. This fluke also causes liver cirrhosis.

5. Examine a prepared slide of a blood fluke (*Schistosoma mansoni*). The blood fluke is an aggressive parasite. People contract this parasite while wading in shallow pools of water. The larva of the blood fluke simply penetrates the flesh and enters the circulatory system. Once in the circulatory system they make their way to the liver. In the liver the larva becomes an adult. This is what you will see in the prepared slide. This particular fluke is especially common in the shallow waters behind the Aswan dam in Egypt. Sketch and note the size and color of the blood fluke.

6. Discuss swimmer's itch. Swimmer's itch is a group of flukes that primarily infect birds. However, people can get this parasite as an accidental host. The swimmer's itch larva penetrates the person's skin similar to the blood fluke. They are not as aggressive as the blood fluke larva. The person's immune system is able to combat the larva as they burrow into the skin. The resulting rash caused by the swimmer's itch is actually the product of your body's immune system killing and dissolving the invading larva.

7. Examine a prepared slide and the preserved specimen of the beef tapeworm (*Taenia saginata*). The beef tapeworm resides in the intestine of cows and pigs and can be very long. People can also get this tapeworm by ingesting undercooked or raw beef infected with the tapeworm larva. This slide contains the head end of the tapeworm. It is called the scolex. Notice the hooks on the scolex and the suckers. These are used to hold the tapeworm in place. One oddity of the tapeworm is that it has no digestive system. It does not require one. The host digests everything that it needs. The tapeworm sits in the intestine and absorbs all the nutrients it requires. Figure 12.3.

Procedure B: Nematoda

1. Make a wet mount of the vinegar eel (*Turbatrix aceti*). The most obvious feature of the vinegar eel is the wiggling motion of thousands of worms. Roundworms only have longitudinal muscles. So their movement is actually a thrashing back and forth through the water. These roundworms are not parasites. They are living on the decomposing fruit within the jar.

2. Examine a prepared slide of the *Trichinella spiralis*. These roundworms cause the disease Trichinosis. People get Trichinosis from eating undercooked pork or bear meat. After eating infected meat that has not been prepared correctly the worms are released into the intestines. The males soon fertilize the females and then they die. The female burrows into the intestinal walls and after a few days release infective larva. Thousands

of these larvas penetrate the circulatory system and lymphatic system. They then move throughout the body in order to burrow into skeletal muscle.   The symptoms of Trichinosis are flu-like when the worms are active. A person will have a slight fever and aching muscles. Occasionally, this disease is fatal. This usually occurs when the heart muscle is infected. Sketch the *Trichinella spiralis*.

3. Examine a prepared slide of a hookworm (*Necator americanus*). The adult form of the hookworm is found in the intestine. At the mouth end of the organism there is a pair of sharp lances that are used to cause a wound in the intestine. This wound bleeds and the hookworm consumes the blood. People contract this disease by ingesting fecally contaminated food. Another type of hookworm is common in cats and dogs. People who do not properly dispose of cat and dog feces run the risk of catching this disease. Sketch the hookworm.

4. Examine a prepared slide of a pinworm (*Enterobius vemicularis*). The pinworm is another parasite that lives in the intestine of people. Specifically, they are usually found in the colon near the appendix. When the females are ready to lay eggs they emerge in the anal region. This causes considerable irritation, which the host itches and scratches and picks up the eggs under the fingernails. If the host fails to wash their hands they can

spread the eggs by simple contact with others. **WASH YOUR HANDS AFTER SCRATCHING....** Sketch the pinworm.

Procedure C: Annelida

1. Study the earthworm model and locate the following structures; mouth, pharynx, esophagus, crop, gizzard, intestine, anus, dorsal blood vessel, ventral blood vessel, hearts, brain, ventral nerve cord, seminal vesicles, seminal receptacles, nephridium, and the clitellum.

2. Obtain a preserved earthworm for dissection. Follow the instructor's procedure for dissection, which will be given during the dissection portion of this lab. Note the function of the structures of the worm mentioned above. (figure 12.4)

    clitellum

    mouth

    pharynx

    esophagus

    crop

    gizzard

    intestine

anus

dorsal blood vessel

ventral blood vessel

hearts

brain-like structure

ventral nerve cord

seminal vesicles

seminal receptacles

nephridium

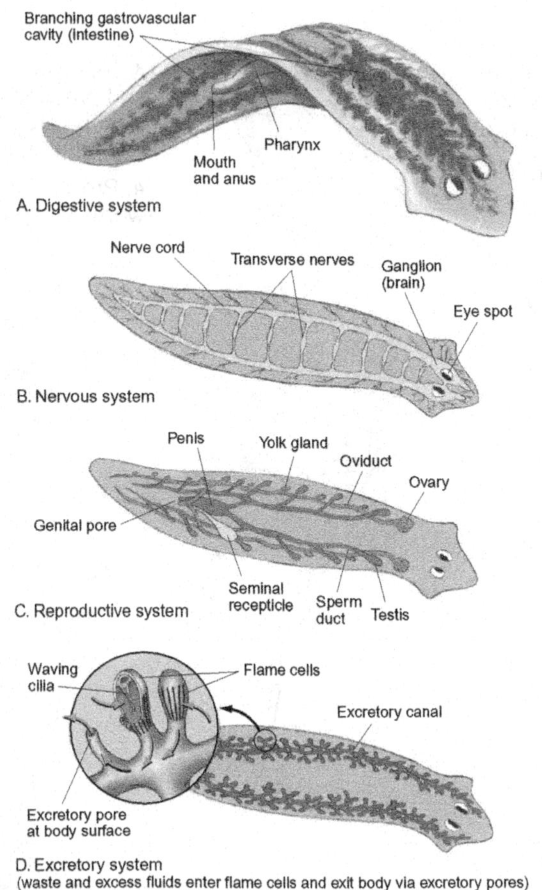

Figure 12.1 Planaria © Image courtesy of Kendall Hunt

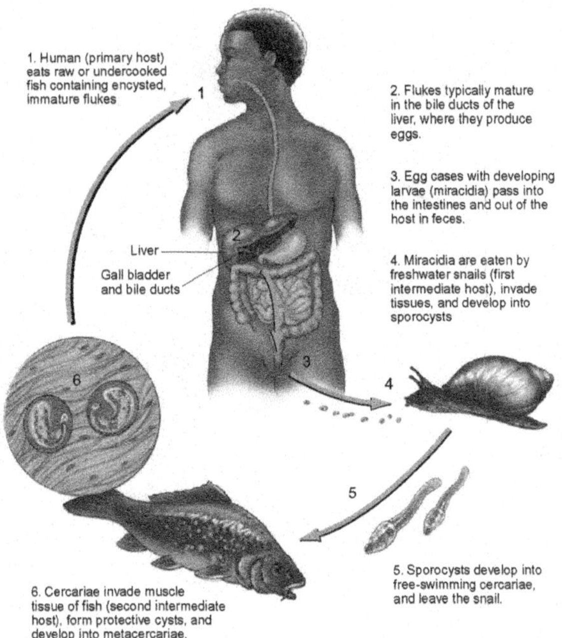

Figure 12.2 Sheep Liver Fluke Life Cycle

© Image courtesy of Kendall Hunt

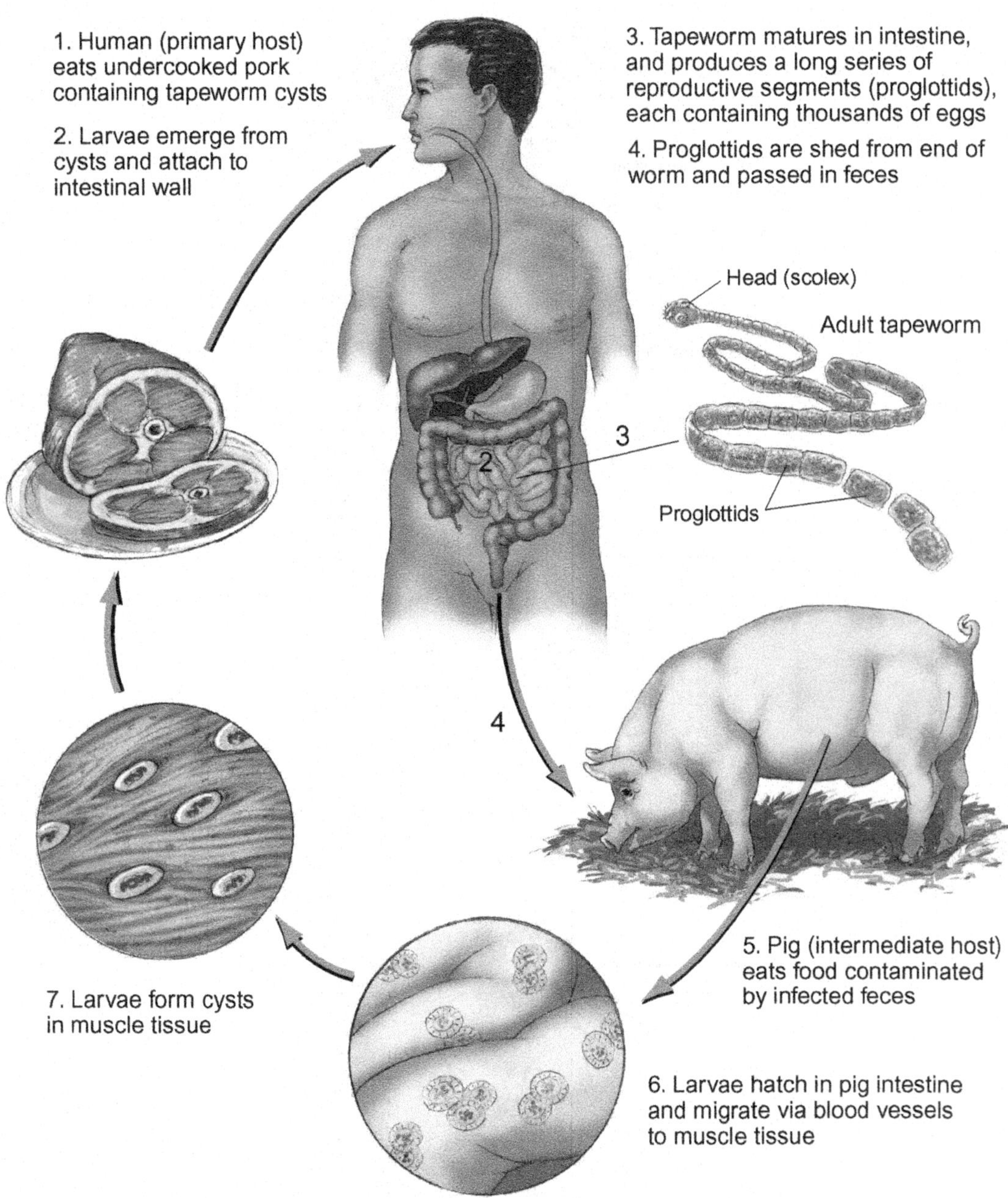

Figure 12.3 Tapeworm Life Cycle

© Image courtesy of Kendall Hunt

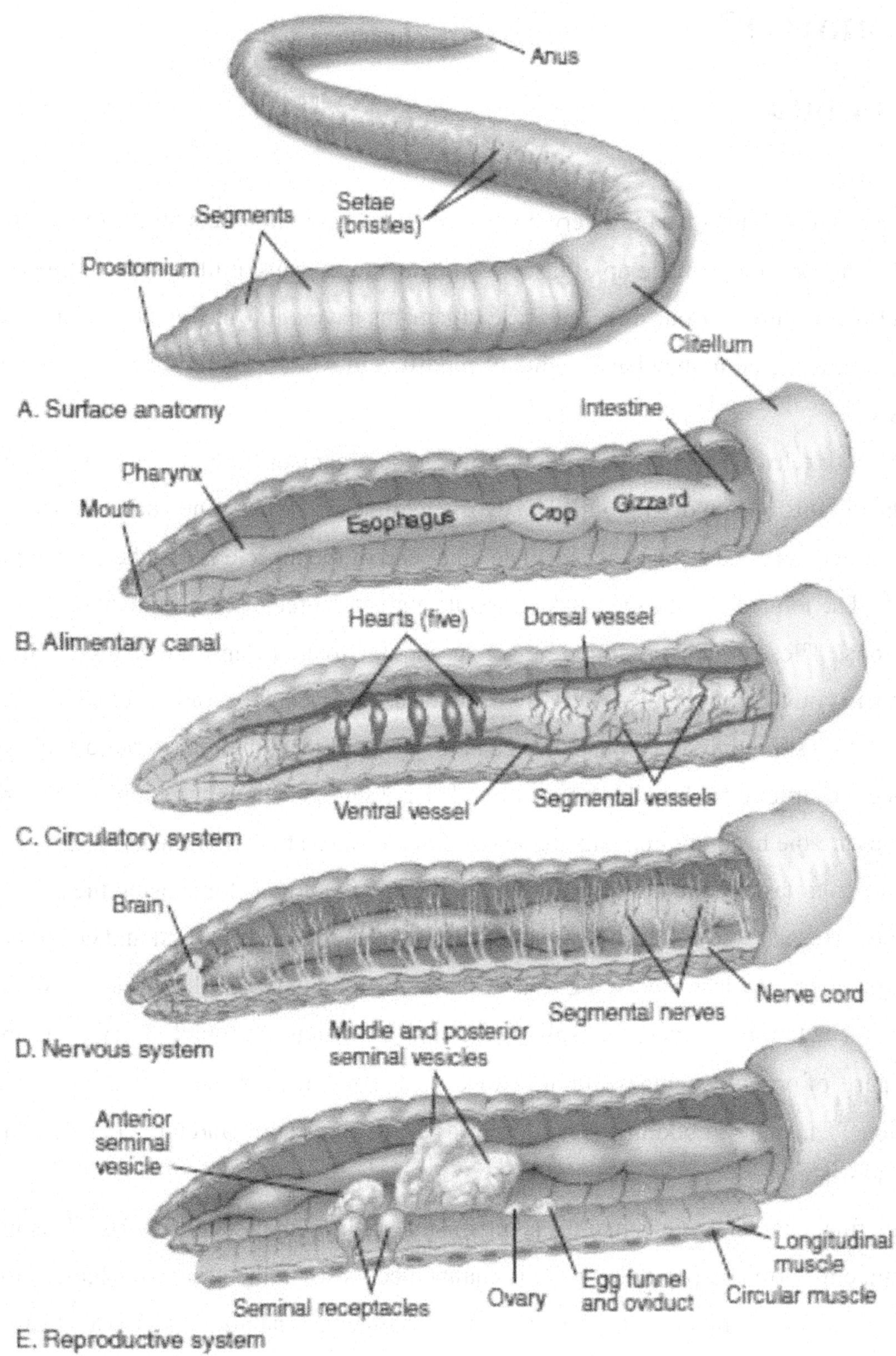

Figure 12.4 Earthworm Anatomy

Laboratory 13

Arthropoda

Introduction

The phylum Arthropoda is a very diverse phylum. It out numbers all the other phyla combined. This phylum is made up of insects, spiders, crustaceans, millipedes, centipedes, and many other lesser-known groups. All of these organisms have a few characteristics in common. They have an exoskeleton; they have segments modified into body regions and they have jointed appendages.

The exoskeleton is for protection. It protects the organism from mechanical injury and it protects the organism from drying out. The exoskeleton is also where one end of muscles that move the jointed appendages is attached. The other end of the muscle is attached to the jointed appendage. The jointed appendages make possible a great degree of environmental manipulations. These organisms are no longer susceptible to environmental change. These organisms can crawl, swim or fly to new environments. They can also pierce, chew, tear, bite, scrape, suck, and sting prey for food. They can also move objects in order to build shelters and other protective structures. Finally, the arthropods have segmented body regions. The three body regions are the head, thorax, and abdomen. Each of these body regions is modified to perform particular tasks. The head is modified to contain the organs for sensing the environment. Because the head is what moves through the environment first and encounters the stimulus first, then the sensor receiving organs should be on the head. The thorax is modified for locomotion. The thorax is where the walking legs and the wings are found. Finally the abdomen has many parts of organ systems within it. However, the abdomen is specialized for reproduction. The male's abdomen is modified for depositing sperm and the female's abdomen is modified for laying eggs.

One final characteristic is that these organisms go through metamorphosis. This is a drastic change in form and physiology. This change occurs when the organism changes from an immature stage, a larva, to an adult. There is also a change in behavior. The larva usually eats different food than the adult and it usually lives in different habitats than the adult. This change in behavior is to alleviate competition between the larva and adult.

In this lab we will look at two examples of the phylum Arthropoda. They are the grasshopper and the crayfish.

Objectives

1. To become familiar with the adaptations of some of the members of the arthropods.
2. To be able to recognize the external structure of the grasshopper and crayfish.
3. To be able to identify the internal structures of the crayfish.

Materials

1. Preserved specimens of grasshoppers and crayfish.
2. Dissecting trays
3. Dissecting tools
4. Crayfish Model

Procedure A: Grasshopper

The grasshoppers that we will use today are the lubber grasshoppers (*Romalea microptera*). These grasshoppers are found in the southern portion of the United States where it is hot and humid most of the year. They live in open fields and feed upon vegetation.

1. Obtain a preserved grasshopper and rinse it thoroughly. Place your grasshopper in the dissecting tray with the dorsal side up. Place it under a dissecting microscope. Examine the external structure.

    The body of the grasshopper like all insects is divided into three body regions; head, thorax, abdomen. Locate these body regions and find the following structures on them. Figure 13.1. The head has two structures you need to find: the antenna, which is used to detect touch and chemicals and the compound eye. The compound eye has many lenses. However, it is still used for sight. The thorax will have the walking legs and wings, which are used for locomotion. The abdomen of the female grasshopper will have an ovipositor. Another structure you need to find is the spiracles. These are located along the side of the thorax and abdomen. Their function is the intake of air.

2. In order to sex your grasshopper you need to observe the end of the abdomen. The female will have two triangular structures that resemble a beak. These are the ovipositors. The male's abdomen is much more round with no protrusions. Another way to tell the sex of the grasshopper is by the relative size of the individual. The males are much smaller than the females.

3. Study the diagram of the internal organs of the grasshopper. Locate the following structures; mouth, esophagus, crop, gizzard, stomach, gastric caeca, intestine, rectum, anus, brain, ventral nerve cord, ganglion, heart, excretory tubules. Figure 13.1

Procedure B: Crayfish

1. Obtain a preserved crayfish from your instructor and place it in the dissecting tray. First you need to find the external structures before you begin dissecting. Figure 13.2. The crayfish has two body regions: the abdomen, which everyone likes to call the tail, and the cephalothorax. The cephalothorax is the head and thorax fused together.

The crayfish has nineteen segments each bearing a single pair of appendages. As we discuss each pair remove the appendage from one side of the crayfish and align them in order in your tray (except the mouth parts). The first pair of appendages on the crayfish is called antennules. These are the short antennas on the head. These function in detecting touch and chemicals. The second pair of appendages is the antennae. These are very long and also function in detecting touch and chemicals. The third, fourth and fifth pairs of appendages are the mouthparts. Each of these appendages has separate names but they are very small and overlapping. The sixth, seventh, and eighth pairs of appendages are the maxillipeds. These are used for gathering food and putting it into the mouth. The ninth pair of appendage is the large pinchers called chelipeds. These are used for defense. The appendages of segment ten through thirteen are called the walking legs. These appendages function for movement. The fourteenth through the eighteenth appendages are the swimmerets. These are for swimming backwards and some are modified as reproductive structures. The last pair of appendages, number nineteen, is the uropods. The uropods are flipper-like and are used for paddles to provide the powerful thrust that is needed to swim backwards.

2. Carefully cut away and remove the dorsal portion of the exoskeleton on the cephalothorax. To do this you must use scissors starting at the junction of the cephalothorax and the abdomen and cut forward towards the eyes. Once you cut as far as you can turn and cut towards the chelipeds. The specific structure that you are cutting is called the carapace. The carapace is not attached along the walking legs or at the junction between the cephalothorax and the abdomen. Next, remove both sides of the carapace. Under the carapace you will find the gills. These resemble wet feathers. The carapace is not attached near walking legs because water needs to flow up over the gills in order for the crayfish to get oxygen. Once you have removed both sides of the carapace locate the following structures. Figure 13.2. Mouth, stomach, midgut, intestine, anus, brain-like structure, ventral nerve cord, segmental ganglia, heart, liver, and green glands.

In order to dissect the abdomen you must carefully cut along the dorsal surface starting at the junction of the cephalothorax and the abdomen, cutting down towards the uropods. Once your incision is to the uropod, peel back the exoskeleton on both sides. At this time you will notice two structures, the intestine which is situated in a grove and the large abdominal muscles.

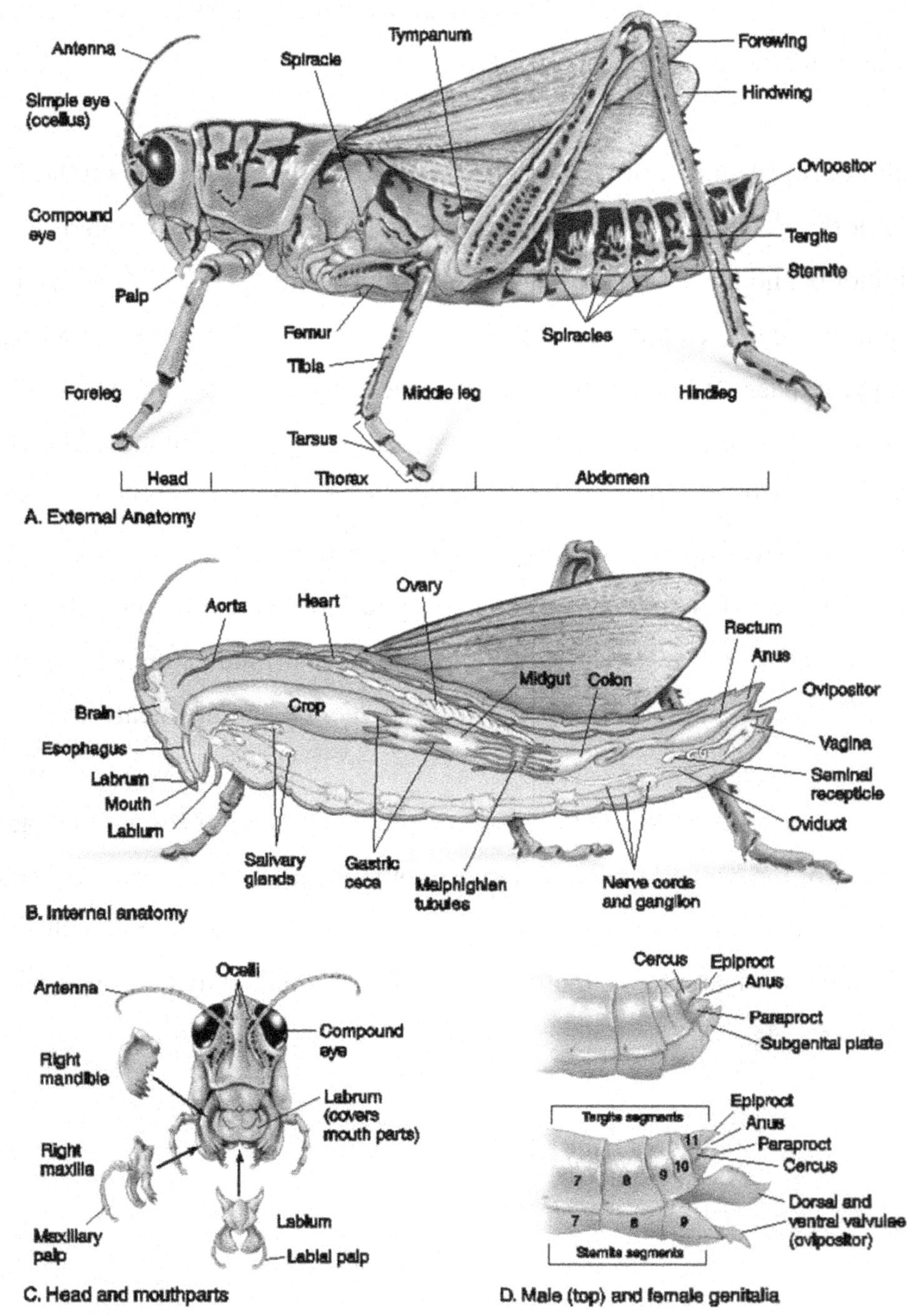

Figure 13.1 Grasshopper Anatomy and Morphology

© Image courtesy of Kendall Hunt

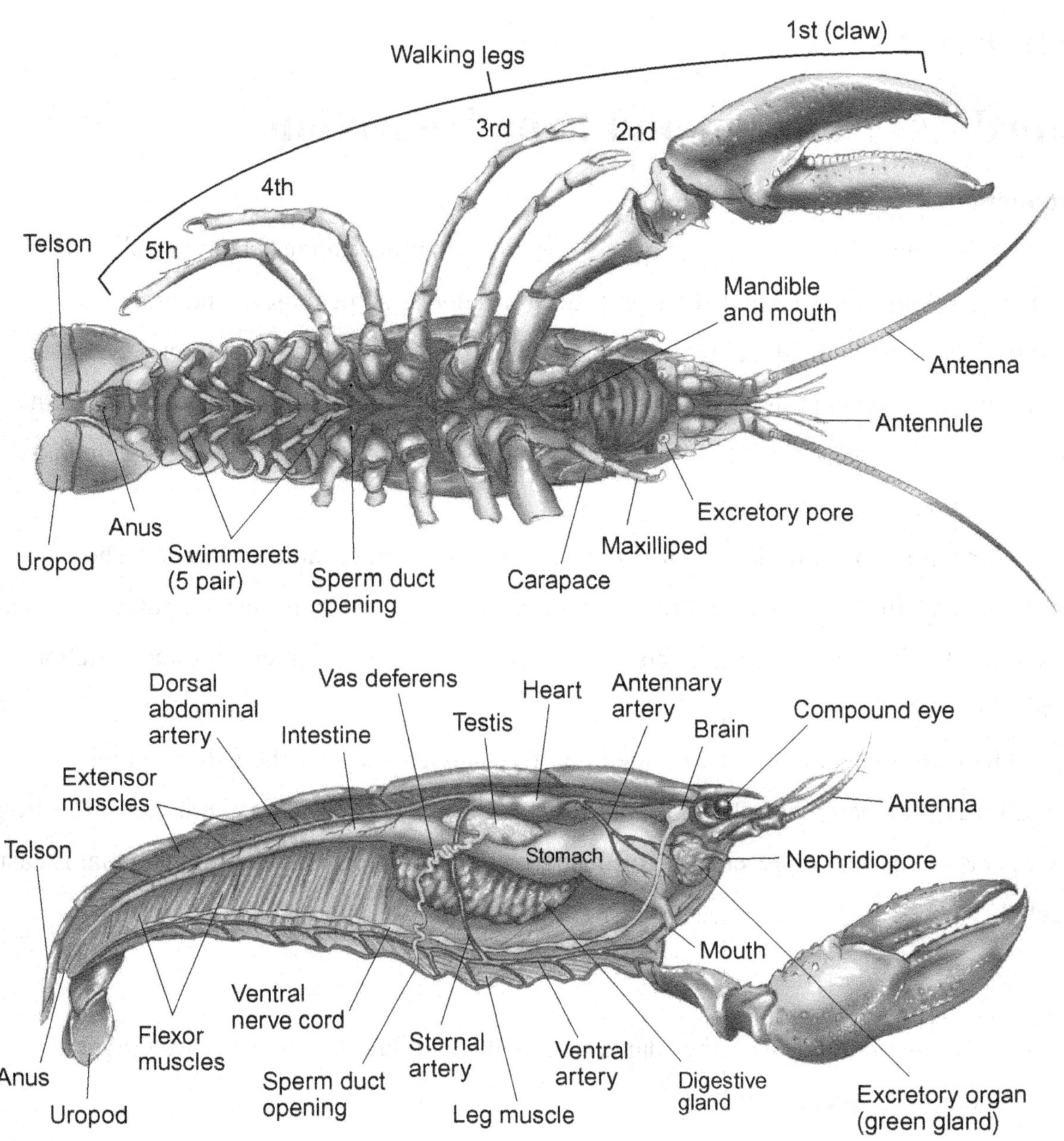

Figure 13.2 Crayfish Anatomy and Morphology

© Image courtesy of Kendall Hunt

Laboratory 14

Chordata: Lancelet and Frog Dissection

Introduction

The phylum Chordata is the most advanced phylum of animals. They are also the most successful phylum. This advancement and success is due to some major structural developments. First is the development of a brain with many brain regions to process more stimuli. The development of an endoskeleton was also very important. The precursor to the endoskeleton is the notochord. The notochord is a dorsal, cartilaginous rod, which extends almost the entire length of the body.

There are two major groups of chordates, the invertebrates and vertebrate. The invertebrates retain the notochord throughout their life. The vertebrates have a notochord as an embryo. By the time the organism is born the notochord has development into the vertebral column.

The vertebral chordates are divided into five classes: they are the fish, amphibians, reptiles, birds, and mammals. The frog that we are going to dissect is an amphibian. The frog is by no means the perfect representative of this phylum. However, it is on organism that is easily dissected.

Objectives
1. To become familiar with the adaptations of the chordates that has allowed them to become very successful.
2. To be able to distinguish between features that makes the vertebrate and the invertebrate chordates similar and different from one another.
3. To be able to identify the external and internal structures of the frog, using diagrams and dissections.

Materials
1. Dissecting trays
2. Dissecting tools
3. Preserved specimens of lancelets

4. Prepared slides of the longitudinal section of the lancelets

5. Preserved specimens of frogs for dissecting

6. Frog skeletons

Procedure A: Lancelets

1. Examine a prepared slide of a lancelet (*Amphioxus sp.*). The lancelet is an organism that is abundant in tropical and temperate seas of the orient. The lancelet is an invertebrate chordate (it has no backbone). This organism retains the notochord throughout its life. Compare the microscope slide to diagram. Figure 14.1. Locate the following structures, notochord, gills, cirri, pharynx, and intestine.

2. Examine the preserved specimen of the lancelet. **Do not open the jar**. Notice the overall body shape of this organism. It has evolved this shape for swimming. The lancelet can swim both forwards and backwards. However, most of the time the lancelet has its tail burrowed into the sandy bottom with its anterior end protruding up into the column of water collecting food.

Procedure B: Frog Dissection

1. Examine the frog skeleton. Notice the relative position of the bones. As you dissect you will need to cut the breastbone (sternum). This is the bone on the ventral surface between the forearms. Next notice the skull. You are going to attempt to dissect the brain. The brain is in the cranial cavity, which is within the bone between the two eyes. It is very small so you must have patience and proceed delicately.

2. Obtain a preserved frog and place it in a dissecting tray. First locate the tympanum. These are round membranes that are located on either side of the head behind the eyes. Their function is to detect sound. Next find the nostrils. The nostrils are for smell. Finally, locate the tongue. The tongue of the frog is attached at the anterior end of the mouth. This allows for a greater extension of the tongue while catching insects.

3. Place the frog in your dissecting tray with the ventral surface up. If the frog's forearms are covering the ventral surface they will have to be removed. With your scissors carefully cut the skin off the ventral surface exposing the abdominal muscles. These muscles must be removed next. To do this, stick one side of your scissors into the muscle near the middle of the abdomen near the hind legs. Cut forward all the way up to the base of the jaw. As you get closer to the top you will begin to cut the sternum. Once you have reached the base of the jaw, cut down and around each side. Now you should be able to remove the muscle. Locate the following structures. Figure 14.2. Liver, stomach, small intestine, large intestine, cloaca, fat tissue, kidneys, adrenal gland, heart, lungs, stomach, spleen, gall bladder, pancreas, urinary bladder. We are not going to be concerned with the reproductive structures. If you have any gray and white material within the abdomen these are eggs and your frog is a female. If your frog does not have any eggs it does not mean that your frog is necessarily a male. It may be a female that is not producing eggs at this time.

4. In order to find the brain you must turn your frog over so the dorsal surface is up. First remove the skin using a scalpel from the head region. Carefully make a longitudinal cut between the eyes through the muscle and bone. Do not make any unnecessary cuts. With the forceps carefully remove the bone and muscle of the skull to reveal the brain. The most obvious part of the frog's brain is the optic lobe. As you move from the optic lobes towards the spinal cord you encounter a "Y" shaped region called the medulla oblongata. This structure regulates the vital organs. Between the top part of the "Y" and the optic lobes lies some tissue called the cerebellum. The cerebellums function is balance. As you go forward from the optic lobes you will encounter the cerebrum. The cerebrums function is memory. Past the cerebrum are the olfactory lobes. The olfactory lobes function is smell. Hand out.

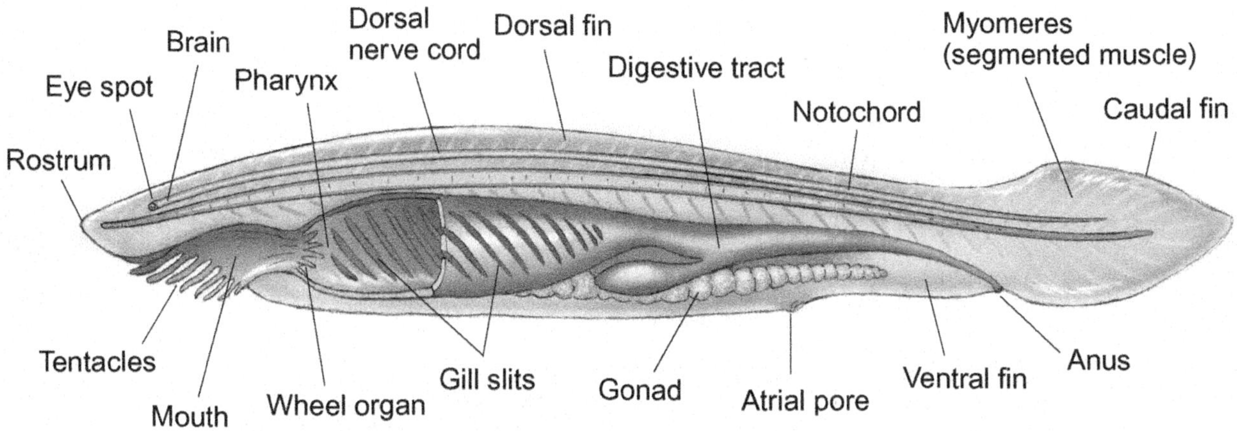

Figure 14.1 Lancelet Anatomy (*Amphioxus sp.*)

© Image courtesy of Kendall Hunt

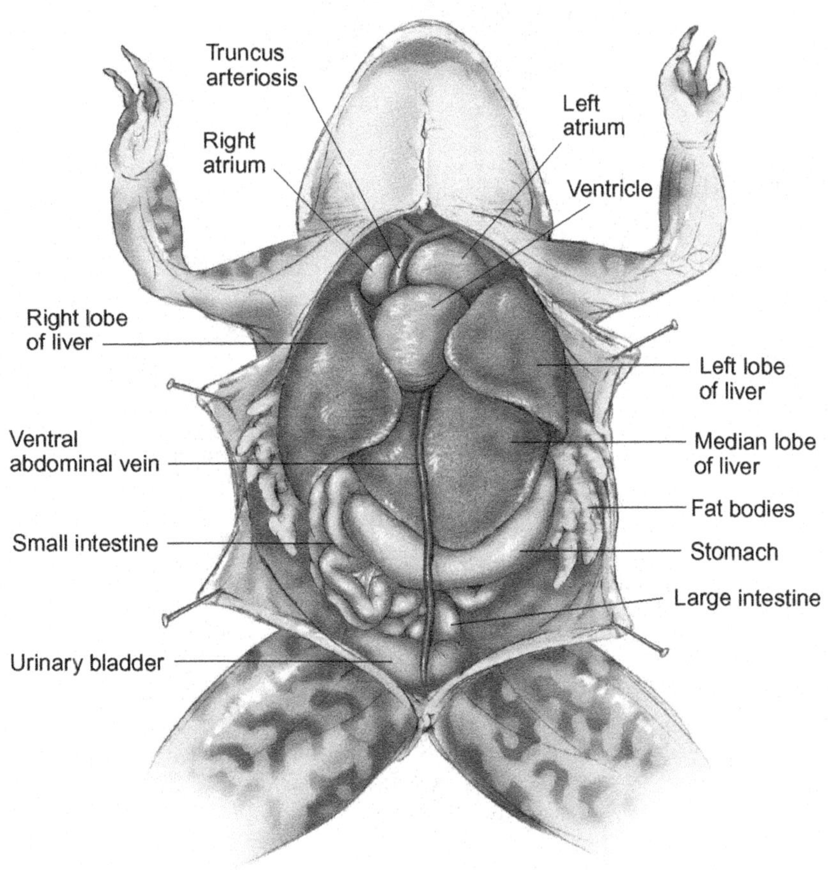

© Image courtesy of Kendall Hunt

Figure 14.2 Frog Anatomy

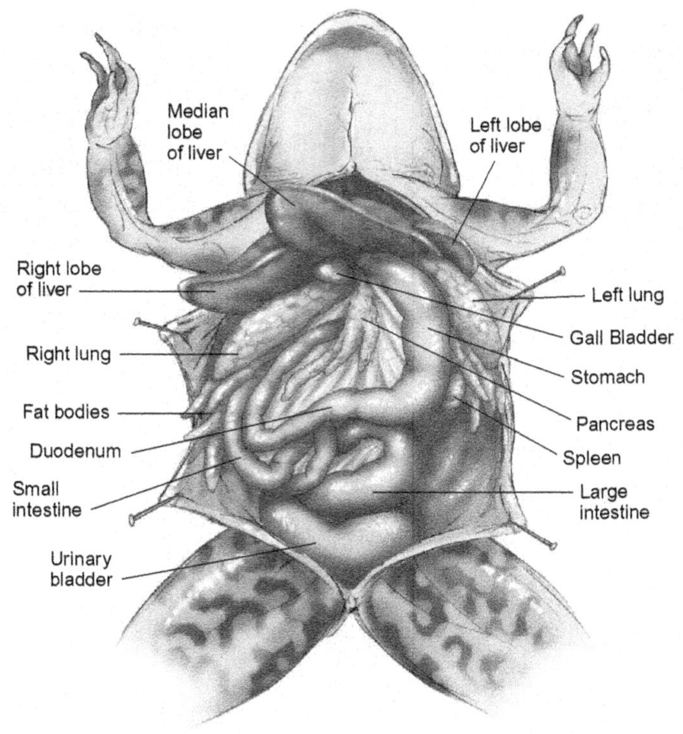

© Image courtesy of Kendall Hunt

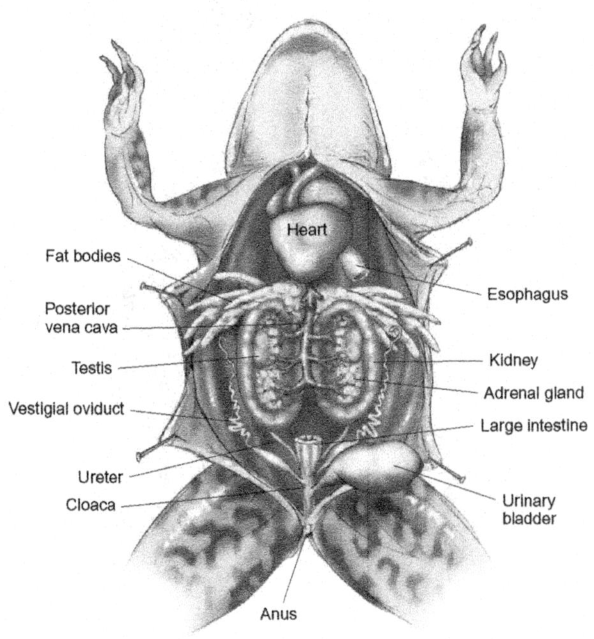

© Image courtesy of Kendall Hunt

Figure 14.3 Frog Anatomy

Laboratory 15

Chordata: Pig Dissection

Introduction

The purpose of this second lab on the phylum Chordata is to study mammalian anatomy and physiology. In this lab we will also examine certain comparative aspects and the relative size and shape of the organs of mammals. The fetal pigs that we are going to look at today were removed from their mothers who were slaughtered for food. The reason why the pregnant pigs were sold for slaughter is because the pigs are sold by the pound and pregnant pigs weigh more than pigs that are not pregnant.

Objectives

1. To be able to identify the external and internal structures of the pig, using diagrams and dissections.

Materials

1. Dissecting trays
2. Dissecting tools
3. Preserved specimens of fetal pigs for dissecting

Procedure A: Pig

For this dissection you need to obtain a fetal pig and place it in the dissecting tray with the ventral surface of the pig up. First, tie a string around one forelimb and place the string around and under the pan and tie it to the other forelimb. Make sure that the forelimbs are spread apart. Next, repeat this procedure for the hind limbs. Remember that dissection does not simply mean cutting up the pig. Instead, it means exposing the organs in order for review before you begin, compare figure 15.1 to the actual pig.

Comparing the pig's skeleton, to our skeleton. The bones are named the same as ours and the bones are in the same relative position as ours. Be able to identify the major bones in both our skeleton and the pig's skeleton.

Observe figure 15.2. Make incision on the pig on the ventral surface following the solid lines that are drawn on the diagram of the pig. The numbers indicate the sequence for making the incisions starting with number one and proceeding to number eight. As you cut be very careful not to cut to deep, it is all right to run the scalpel over your incision more than once in order to cut all the way through the pig's skin and muscles. Be careful not to cut any of the pig's organs.

Observe figure 15.3 and find the following organs of the digestive system; stomach, pancreas liver, gull bladder, small intestine, colon (large intestine), rectum and the anus. Next, examine figure 15.4, the pig's respiratory system and parts of the circulatory system. Be able to find the larynx, trachea, bronchus, lungs, plural cavity, diaphragm, heart and aortic arch. Finally, investigate the urogenital system of the pig (figure 15.5 and figure 15.6). Be able to recognize the kidney, adrenal gland, urinary bladder and ureter of either the female or male. Also be able to identify the ovary in a female and the testis in a male.

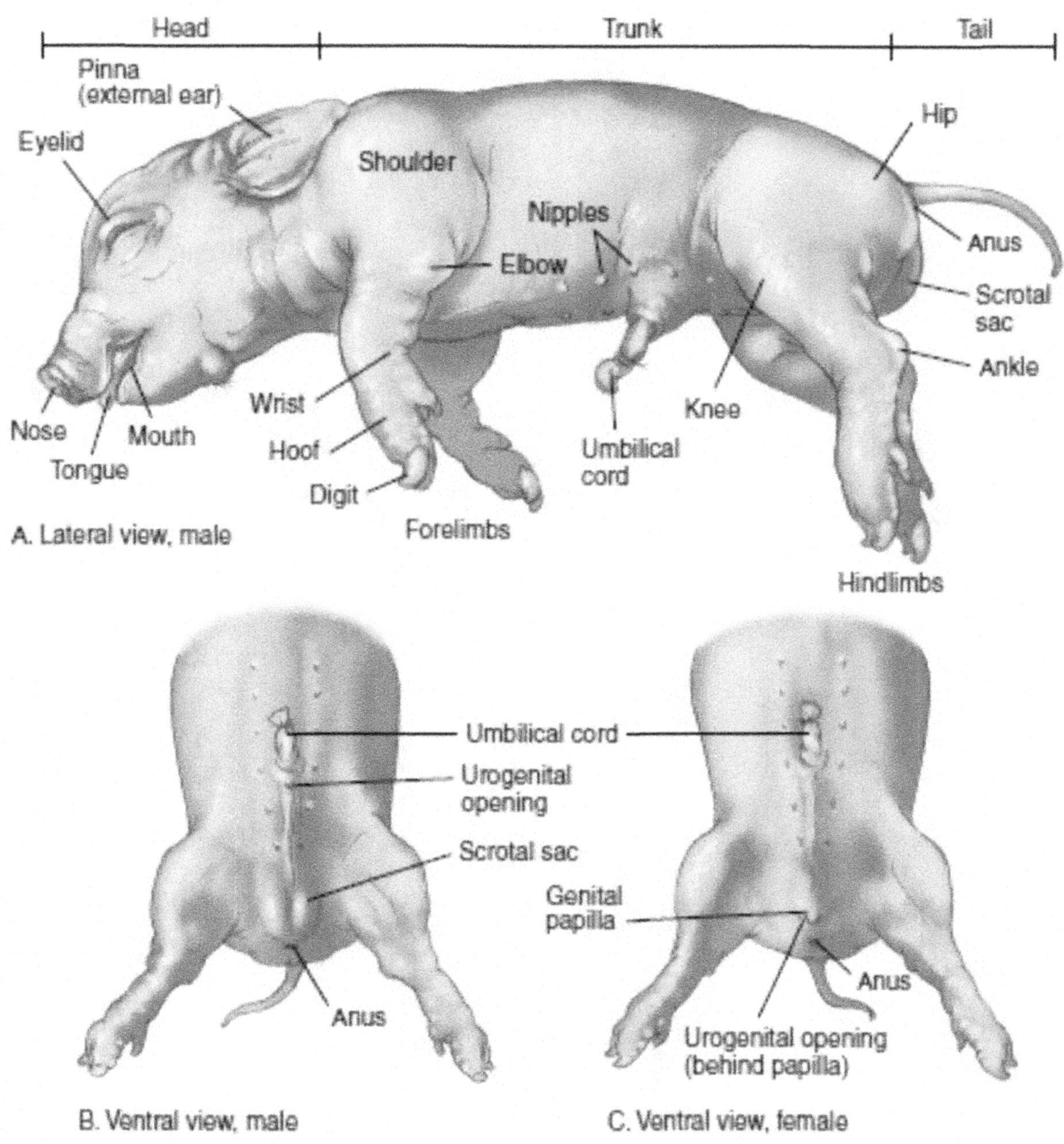

Figure 15.1 Fetal Pig Surface Anatomy

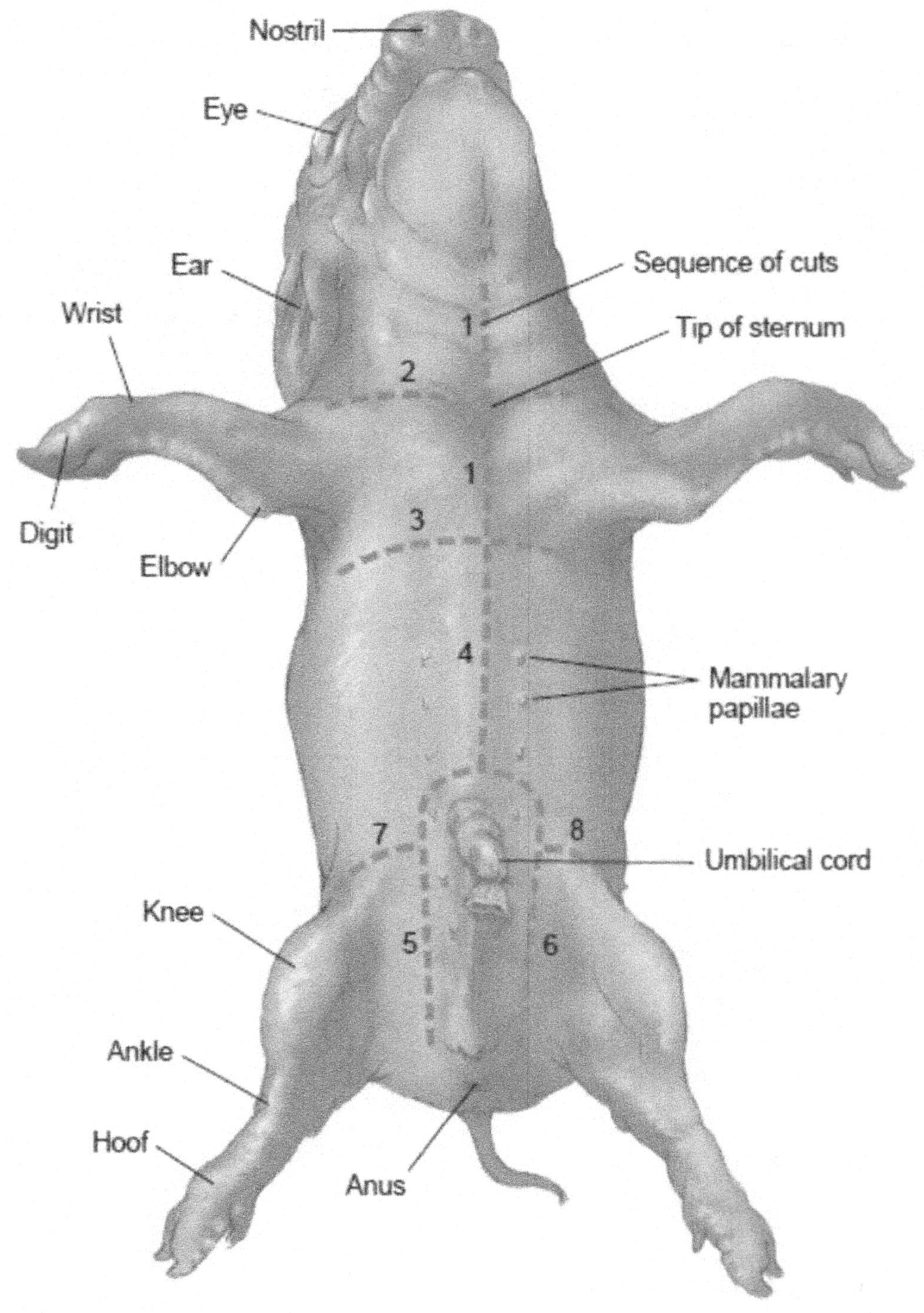

Figure 15.2 Diagram for make Incisions on the Fetal Pig

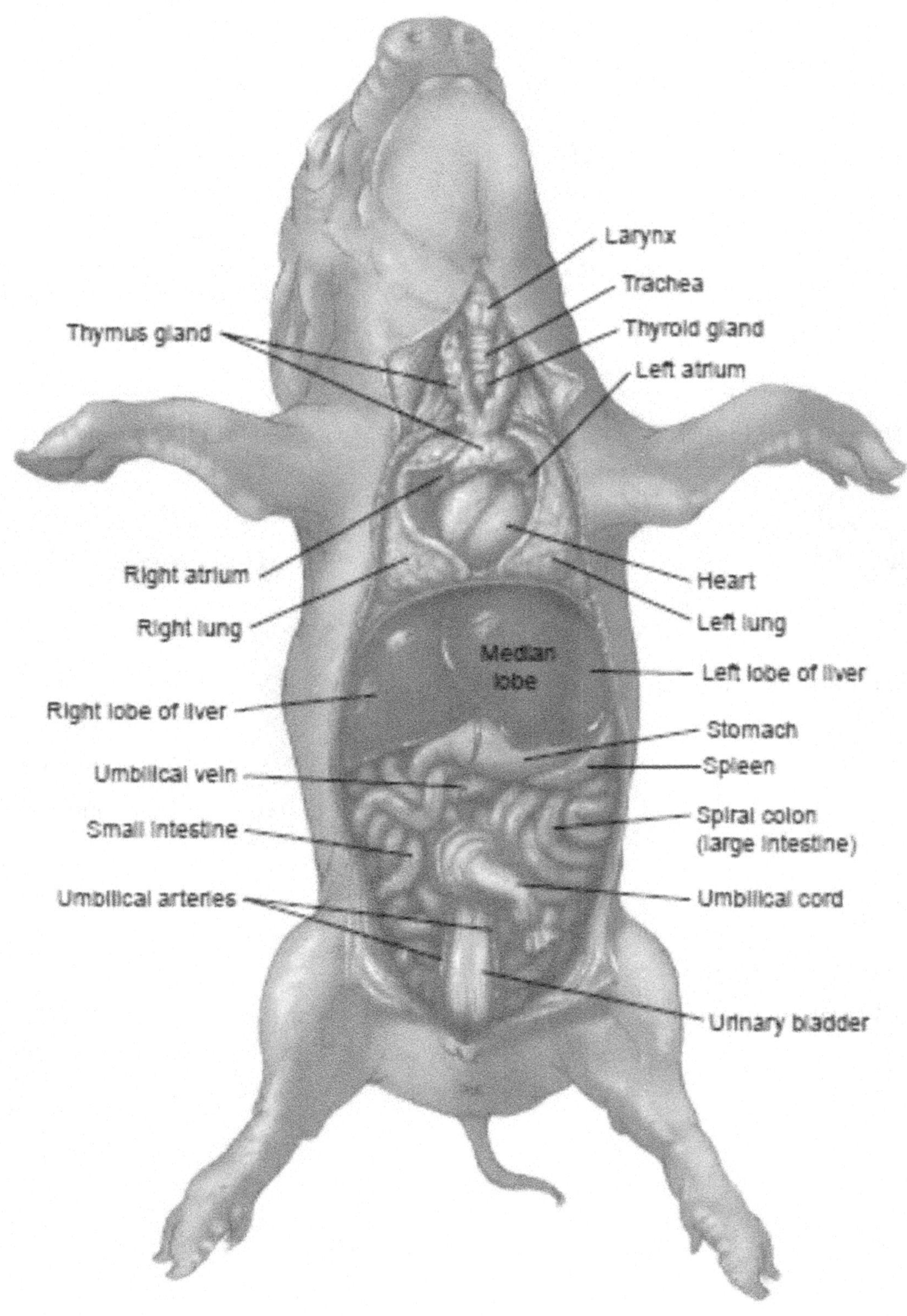

Figure 15.3 Major Organ Systems

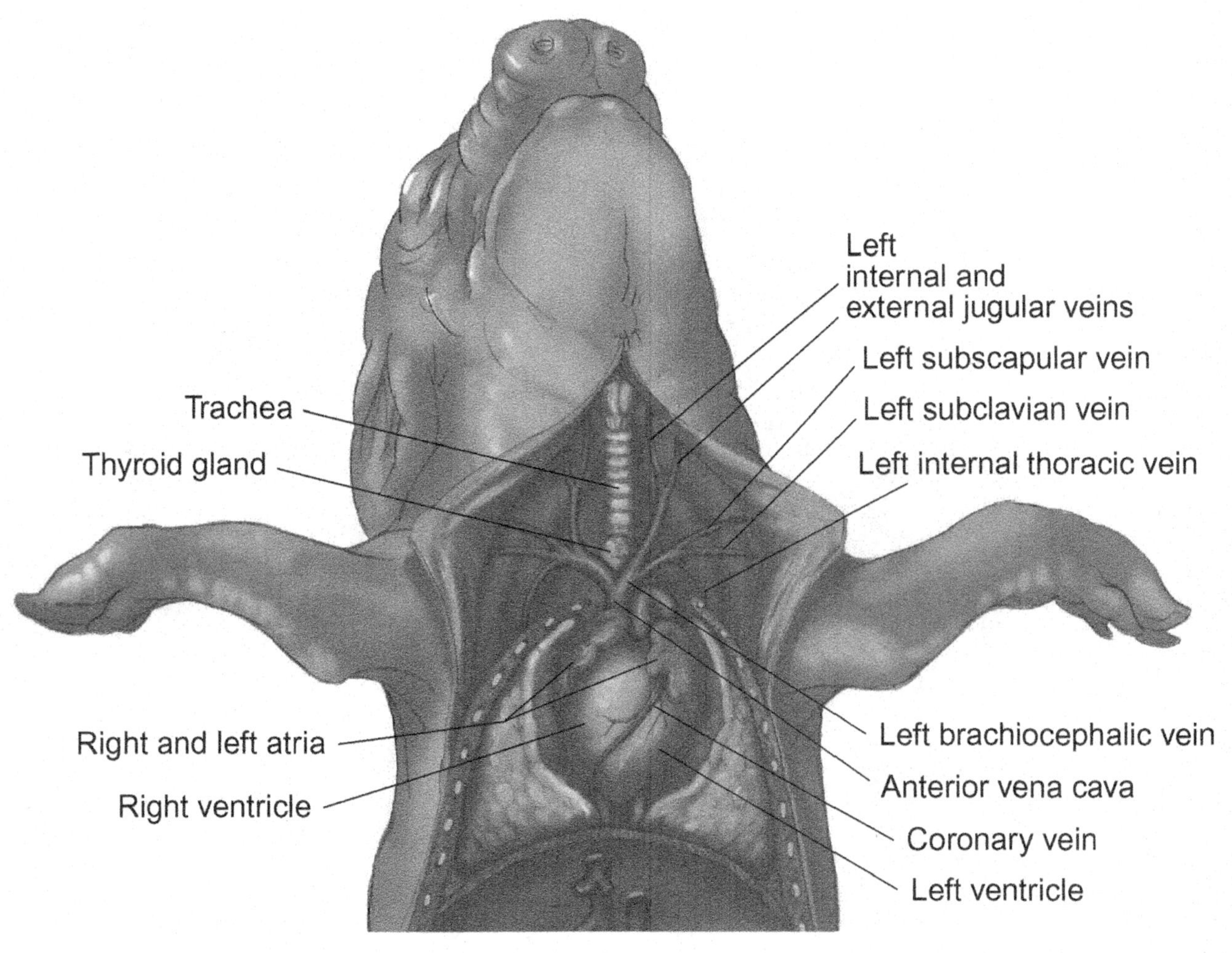

© Image courtesy of Kendall Hunt

Figure 15.4 Fetal Pig Respiratory System and Circulatory System

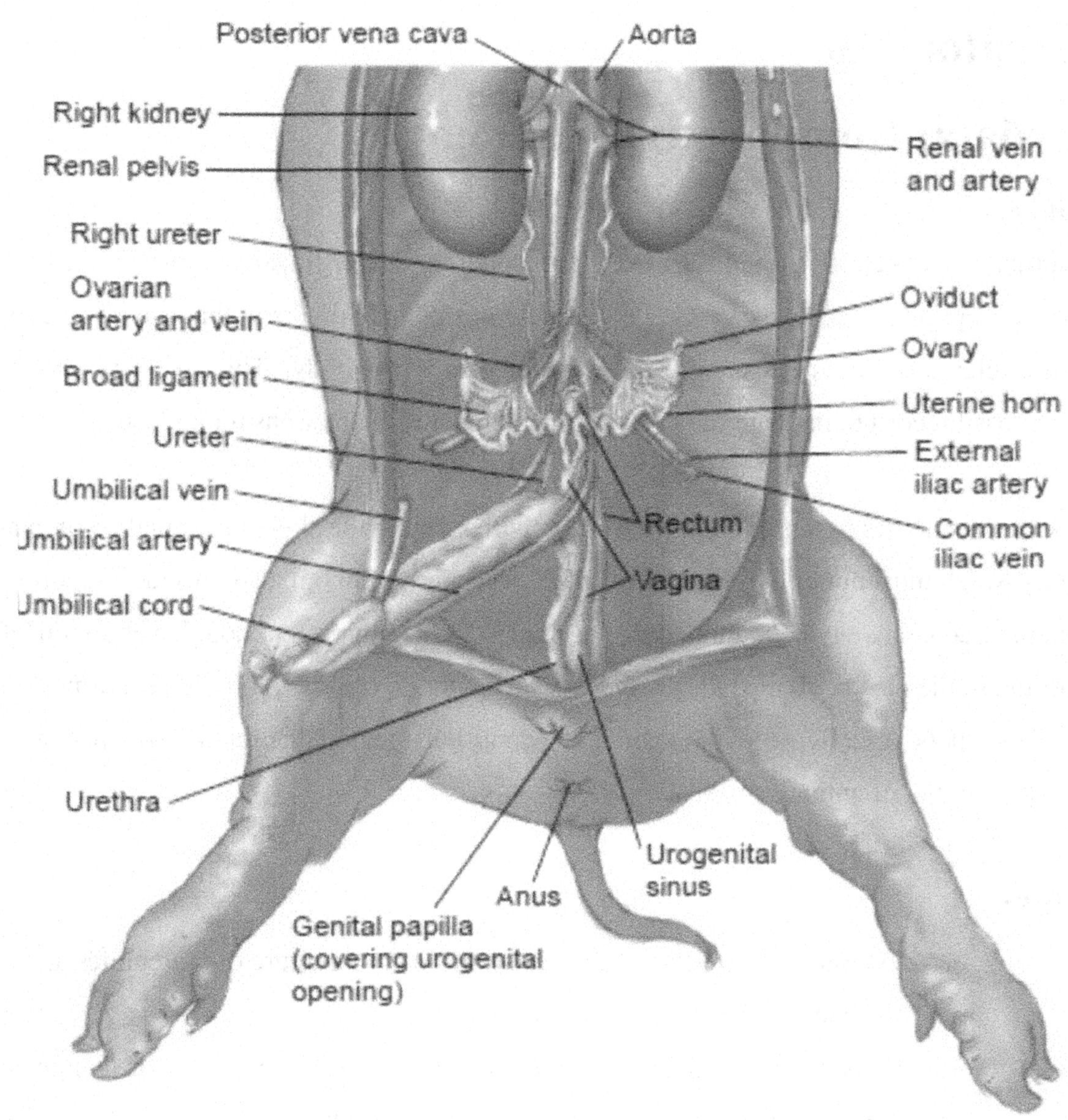

© Image courtesy of Kendall Hunt

Figure 15.5 Female Fetal Pigs Urogenital System

Laboratory 16

Chordata: Comparative Anatomy

Introduction

The chordate endoskeleton provides an excellent model for studies in adaptation. There are many common traits which we can see, yet there is considerable diversity that can also be found. The endoskeleton is more permanent than the softer parts of the body. Thus, we have an excellent fossil record to indicate the manner in which many adaptations took place.

There are three themes in this exercise. First, we will examine the skull and spinal column as structures which must not only protect the central nervous system but also provide support and movement. Second, we will examine the appendages and see that even though a common basic form occurs in the design of appendages there is sufficient flexibility to provide for arms, legs, wings and flippers. Finally, we will examine the dentition and note the relationship between the jaws, teeth and diet of animals.

Objectives

1. To develop an awareness of the variations in the structure of vertebrates and relationships between structure and function.
2. Identify some of the major differences between the jawless fish, cartilaginous fish, boney fish, amphibians, reptiles, and mammals.
3. Identify the following structures of the jaw: maxilla, mandible, condyle and fossa.
4. Recognize the following structures of a tooth: root, crown, cusps, dentine, enamel, and pulp.
5. Identify the following kinds of teeth by appearance and function: incisor, canine, premolar, and molar.
6. Identify the sagittal crest and zygomatic arch.
7. Recognize the major differences between insectivores, carnivores, herbivores, and rodents.
8. Observe and be able to indicate whether a given skull is insectivores, carnivores, herbivores, or rodents.

Materials

1. Preserved specimens of the following: lamprey, shark, shark skeleton, bony fish, mud puppy, crocodile, snake, turtle, cat, and bat.

2. Skulls of insectivores, carnivores, herbivores, and rodents.

Procedure A: Jawless Fish

Observe the preserved lamprey, this is a primitive chordate. The lamprey lack jaws, bones and paired appendages. Even though the lamprey has no jaws it does have many very sharp thorny teeth. The lamprey also has a skeleton; however, it is made of cartilage, not bone, like our nose.

Procedure B: Cartilaginous Fish

Examine the skull of the shark in the specimen jar. The skull and the rest of the skeleton are composed of cartilage which never hardens into bone throughout the life of the shark. If the shark is so soft why is it feared as such a ferocious animal?

Examine a shark specimen. Note the jaws and teeth. Teeth are perhaps not the best term for these structures since they are so different from other teeth. We shall call them denticles. They occur in rows and are replaced throughout the life of the shark. The denticles are believed to be the same origin as the scales on the body of the shark.

Procedure C: Bony Fish

Compare the vertebrae at two or three points along the spinal column of the fish. Are there many variations in the size or shape of the vertebrae? Note the paired appendages. Examine the fish skull and observe the kind and distribution of the teeth. The teeth occur not only along the ridge of the jaw but also on the roof of the mouth. The teeth lack specialization, they are all alike.

Procedure D: Amphibian

Examine the model of the mud puppy. Compare the spinal column and appendages with the fish. Can you find mud puppies around here?

Procedure E: Reptile

1. Crocodile - Look carefully at the vertebrae of the crocodile. Although they lack the
    degree of variation which occurs in man, they have considerably more variation than fish.
    The skull has a palate separating the nasal passages from the mouth. This permits
    simultaneous eating and breathing. The teeth are more variable than those of other
    animals we have looked at thus far, but they are still somewhat alike. The term for this
    condition is homodont (like teeth). The appendages are lateral rather than underneath the
    body as they are in mammals.

2. Snake - Look at the teeth on the snake skeleton. Snakes eat their food whole. The
    recurved teeth prevent the prey from escaping but permit swallowing. Note the lack of
    appendages and pectoral or pelvic girdles. The python has a vestige of the hind
    appendages.

3. Turtle - The turtle represents another extreme in adaptation of the skeleton. The shell of
    the turtle consists of fused plates which develop from the skin. There are no ribs and the
    vertebrae are fused to the shell. The shell has taken on some of the functions of the
    endoskeleton, it supports and protects. Although the turtle is lacking teeth it retains an
    effective bite.

Procedure F: Mammal

1. Cat - Examine the vertebrae of the cat. Compare the vertebrae in the neck, torso and tail
    regions. Note the variations in the size of the vertebrae and the size of the spines (vertical
    projections) and processes (lateral projection) of each vertebra. The spines and processes
    provide attachment for muscles and ligaments. The size of the spines or processes
    usually reflects the size of the muscle that is attached to it and the amount of stress which
    is put upon it.

Note the manner in which the pectoral and pelvic girdles are attached to the body and
how they relate to the spinal column. The position of the appendages under the body

rather than lateral is more effective for terrestrial locomotion. Compare your arm with the foreleg of the cat.

Examine the skull of the cat. The brain case is proportionately larger than other animals we have examined so far. Note the jaws and teeth. The teeth vary in size and shape; they are called heterodont (different teeth).

2. Bat - Compare the wings of the bat with your arm. Determine which bones in the bats wings compare with the bones of your fingers, wrist, forearm, and upper arm. Compare the bat's wing to your arm.

Examine the teeth of the bat. This bat lives on insects and the size and shape of the teeth reflect its diet.

Procedure G: Dentition

The skeleton of an animal reflects the manner in which that animal adapts to the environment. The skeleton determines the kinds of movements an animal can make. This is influenced by the size of the bones, the size of the muscle attachment, the amount of leverage, the type of joint and other variables. To a large extent these factors are under genetic control but they may also be influenced by the environment. Consider the size of a given bone, its ultimate potential size may be genetically controlled but diet and use can also influence its growth. Bone growth may be stimulated by stresses and use of the given muscles and bones may increase their strength and size. Lack of use may also result in degeneration of bone tissue. In this exercise we will examine those structures related to dentition (teeth) as examples of adaptation. This will include examining the teeth, jaws and parts of the skull of various animals.

The jaws consist of the maxilla above and the mandible below. The mandible consists of a single bone in mammals. There are three major muscles which are involved in closing the jaw. These muscles attach at different points on the mandible and the skull. The three muscles are the masseter, temporal and pterygoid. Study the diagram of the carnivore skull. The stippled area represents the points of attachment of the different muscles. The size of the different muscles varies in animals with different diets. The size and strength of the various points of attachment will also vary.

The mandible hinges with the skull. The mandible has a rounded portion, the condyle, and the skull has a somewhat hollowed area, the fossa. The condyle articulates with the fossa and is held in place by ligaments. The size, shape and position of the condyle and fossa vary according to the diet of the different animals. This joint is somewhat loosely hinged to permit not only opening and closing movement of the mandible but also lateral motion and forward and backward motion. The degree to which these various types of motion occur varies.

Teeth probably evolved from denticles, such as those that occur in the shark. The teeth arise separately from the jaw. They also attach to the jaw in different ways in different animals. Reptiles and mammals both have sockets in the jaw which hold the teeth in place. The portion of the tooth which fits in the socket is the root. Teeth may have one or more roots and the root is held in the sockets by cement. The anterior teeth have a single root while the posterior teeth have multiple roots. The part of the tooth which rises above the gum is the crown. The crown may have one or more rounded projections called cusps. The boney part of the tooth is the dentine. The crown is covered with a thin layer of enamel which is much harder than the dentine. Teeth are replaced continuously in most vertebrates. However, mammals usually have only two sets.

Dentition in mammals is more complex. This is probably to provide greater efficiency in processing food for warm blooded animals with higher metabolic rates. There are four kinds of teeth which may occur in mammals. The incisors may be simple spikes of blades. They are relatively small with a single root. Their function is usually for securing food by biting. Canines are effective for capturing, holding and piercing prey. They may also be used for tearing flesh. There are also two types of molars, the pre-molars and molars. There are usually two sets of pre-molars but only one set of molars. Since they are so similar some simply call them cheek teeth. These pre-molars and molars usually have a broad surface on the crown with cusps arising for the surface. This creates an irregular surface for grinding and crushing food materials.

The number and kinds of teeth vary from animal to animal and may be expressed by a dental formula. In this formula half of the upper and lower jaw is described. If we multiply by two, it will give the total number of teeth. As an example the dental formula for a person is:

$$\frac{\text{Upper jaw} \quad 2 \text{ incisors, 1 canine, 2 premolars, 3 molars}}{\text{Lower jaw} \quad 2 \text{ incisors, 1 canine, 2 premolars, 3 molars}} \; X\,2 = 32$$

We shall abbreviate this formula as:

$$\frac{2-1-2-3}{2-1-2-3} \; X\,2 = 32$$

Most animals are in a process of change because of a perpetually changing environment. This is particularly true of diet. An animal may evolve one type of dentition to suit a particular type of food but the food source becomes diminished and the animals must change its diet or become extinct. For this and other reasons it is difficult to rigidly categorize animals by dentition. However, we shall attempt to categorize in this exercise. Do not be surprised if some animals do not fit in any category. These are broad generalizations.

1. Insectivores – Insects provide the primary source of food for several mammals. Most insects require little or no chewing. The primary problem is capturing and holding the insect until it can be swallowed.

   Some mammals such as the anteater have solved the problem with their tongue. The tongue of the anteater is very long and sticky to capture and hold ants without the use of teeth. The anteater has no teeth.

   Insectivores such as moles do have teeth. However, the teeth are usually small with a low crown but very sharp and all look alike. The sharp teeth are necessary for capturing and holding the insects. Chewing is not necessary for the mole. Strong jaw muscles are not needed for eating insects.

The dental formula of common insectivore, the mole, is:

$$\frac{3-1-4-3}{3-1-4-3} \times 2 = 44$$

2. Carnivores – Capturing, holding and devouring a flesh animal requires specialization of teeth and strong jaw muscles. Carnivores use primarily the temporal and masseter muscles for closing the jaws. Since the temporal muscle attaches to the side and top of the skull this area is well developed. Sometimes a ridge of bone, the sagittal crest, may occur along the mid dorsal line of the skull for temporal muscle attachment. The zygomatic arch is also well developed for the attachment of the masseter muscle.

The most predominant teeth are the canines. These function almost like daggers. They are long and sharp, thus effective for creating puncture wounds or for tearing flesh. The incisors are sharp for cutting. The digestive system of carnivores is adapted to receiving and digesting large chunks of flesh. Meat does not need to be thoroughly chewed. The premolars and molars are laterally flattened like a blade for cutting rather than grinding. In some carnivores such as the dog the upper fourth pre-molar and the lower first molar form what we call the carnassial teeth. The principle function of these teeth is shearing.

The jaws of most carnivores have little lateral motion. The long canines do not permit much lateral movement. This type of motion is not necessary since there is little grinding action required by the pre-molars and molars.

The dental formula of a common carnivore, the cat, is:

$$\frac{3-1-4-2}{3-1-4-2} \times 2 = 40$$

3. Herbivores – Obtaining and chewing herbaceous foods requires some unique adaptations of jaws and teeth. Most herbivores have an elongated snout to better enable them to reach grass, leaves, etc. This results in a jaw which is so long that there are not enough teeth to fill it. A space, the diastema, occurs between the anterior teeth and the pre-molars and molars. This space provides a food storage area. Incisor teeth are flattened

for biting off herbs. In some herbivores the ridge of the bone on the upper jaw functions as teeth and the upper incisors are lacking. Not only are they not needed but long canines would make lateral jaw motion difficult.

Food that is vegetation usually has less energy yield than meat. In addition plants contain a high proportion of cellulose, a substance which vertebrates cannot digest. The teeth, jaws and gut of the herbivores have adapted to these circumstances. Herbivores consume large quantities of food and strong jaws with large masseter and pterygoid muscle are necessary to do extensive chewing. The masseter muscle attaches to the suborbital area of the skull rather than the zygomatic arch. The pterygoid muscle provides the lateral motion necessary for grinding the food.

The per-molar and molar teeth are uniquely adapted for grinding food. They are broad and somewhat flattened and instead of having round cusps they have ridges of enamel. The lateral grinding motion of the jaw causes the teeth to wear. Because the teeth of herbivores wear, they are high crowned.

The digestive system of herbivores is usually much larger than carnivores to permit greater digestive activity. This also permits more microbial activity. Some of these microbes digest cellulose.

The dental formula of common herbivores:

$$\frac{0-0-3-3}{4-0-3-3} \text{ X } 2 = 32$$

4. Rodents – Gnawing or grinding of wood is probably used as much for acquiring shelter as it is for chewing food. Rodents are noted for their two pairs of large incisors. The chisel-like shape is made possible by irregular wear. Only the outer surface is covered with enamel. The softer inner portion wears faster and the incisors are continuously growing.

Rodents are lacking canines. A large diastema and large cheeks provide room for food storage. The pre-molars and molars are flattened and have enamel ridge on the grinding surface. The jaws are hinged to permit the lower jaw to slide back and forth.

The dental formula of common rodent is:

$$\frac{1-0-0-5}{1-0-0-5} \ \text{X}\,2 = 24$$

Procedure H: Example Dental Formulas for Common Animals

A number of skulls have been placed on the front table. Using the following dental formulas, identify and name the skulls.

1. Opossum $\qquad\qquad \dfrac{5-1-3-4}{4-1-4-3}\ \text{X}\,2 = 50$

2. Star nosed mole $\qquad \dfrac{3-1-4-3}{3-1-4-3}\ \text{X}\,2 = 44$

3. Dog, fox, wolf, coyote, bear $\qquad \dfrac{3-1-4-2}{3-1-4-3}\ \text{X}\,2 = 42$

4. Raccoon $\qquad\qquad \dfrac{3-1-4-2}{3-1-4-2}\ \text{X}\,2 = 40$

5. Fisher, marten, wolverine $\qquad \dfrac{3-1-4-1}{3-1-4-2}\ \text{X}\,2 = 38$

6. Otter $\qquad\qquad \dfrac{3-1-4-1}{3-1-3-2}\ \text{X}\,2 = 36$

7. Badger, mink, skunk weasel $\qquad \dfrac{3-1-3-1}{3-1-3-2}\ \text{X}\,2 = 34$

8. Elk $\qquad\qquad \dfrac{0-1-3-3}{3-1-3-3}\ \text{X}\,2 = 34$

9. Shrew $\qquad\qquad \dfrac{3-1-3-3}{1-1-1-3}\ \text{X}\,2 = 32$

10. Bison, cattle, deer, moose
$$\frac{0-0-3-3}{3-1-3-3} \; X\,2 = 32$$

11. Cougar
$$\frac{3-1-3-1}{3-1-2-1} \; X\,2 = 30$$

12. Bobcat, lynx
$$\frac{3-1-2-1}{3-1-2-1} \; X\,2 = 28$$

13. Rabbit
$$\frac{2-0-3-3}{1-0-2-3} \; X\,2 = 28$$

14. Least chipmunk, ground squirrel, woodchuck
$$\frac{1-0-1-3}{1-0-1-3} \; X\,2 = 22$$

15. Squirrel
$$\frac{1-0-2\,or\,1-3}{1-0-1-3} \; X\,2 = 22 \; or \; 20$$

16. Beaver, eastern chipmunk, porcupine
$$\frac{1-0-1-3}{1-0-1-3} \; X\,2 = 20$$

17. Mice, rats, vole
$$\frac{1-0-0-3}{1-0-0-3} \; X\,2 = 16$$

Laboratory 1

End of Lab Questions

Name ____   Lab Section ______________

1. What is the total magnification when you are using Low Power, High Power and the Oil immersion objective lenses?

2. Why is the stage called a mechanical stage?

3. What are the two ways of adjusting the light in the compound light microscope?

4. The fine focusing knob is the ______________ of the two knobs; the coarse focusing knob is the ______________ of the two knobs.

5. When cleaning the microscope, you should use ________________________.

6. Because of the optical illusion, how would this "e" appear?

7. Matching

   a. ______ Wet Mount   1. Largest magnification

   b. ______ Ocular   2. When you look in the microscope

   c. ______ Oil immersion   3. Controls light intensity

   d. ______ Low Power   4. Lens closest to the eye

   e. ______ Field of View   5. Least magnification

   f. ______ Iris Diaphragm   6. Microscope slide

Laboratory 2

End of Lab Questions

Name _________________________________   Lab Section _____________

1. What shape are coccus bacteria? _____________________

2. What shape are bacillus bacteria? _____________________

3. What shape are spirillum bacteria? _____________________

4. Describe how the prokaryotic bacteria are different from the eukaryotic protists.

5. When considering the protozoa what dictates reproducing asexually vs. sexually?

6. What is cytoplasmic streaming?

7. Matching

a. ______ Chrysophyta	1. Red Algae
b. ______ Euglenophyta	2. *Paramecium*
c. ______ Chlorophyta	3. Slime Mold
d. ______ Phaeophyta	4. *Amoeba*
e. ______ Rhodophyta	5. Brown Algae
f. ______ Mastigina	6. *Euglena*
g. ______ Sarcodina	7. *Trypanosoma*
h. ______ Ciliophora	8. *Spirogyra*
i. ______ Myxomycota	9. Diatoms

Laboratory 3

End of Lab Questions

Name ________________________________   Lab Section ______________

1. What is the technical name for the filaments that grow out of the spore? ________

2. Name the fungus that grows on the surface of lilac leaves. ________________

3. Specifically, what are the cells on the inside of the sporangium called? ________

4. Describe the differences between the fruiting body and the fungal body.

5. Which of the following is not used for dispersal by a fungus?
   a. Basidiospore
   b. Ascospore
   c. Zygospore

6. Draw and name the four main structures of the mushroom?

7. Matching

   a. _____ Lichen   1. Zygomycota

   b. _____ Morel   2. Ascomycota

   c. _____ Bread Mold   3. Basidiomycota

   d. _____ *Coprinus*   4. Mycophycophyta

   e _____ *Rhizopus*

   f. _____ Yeast

   g. _____ *Peziza*

   h. _____ Corn Smut

   i. _____ Powdery Mildew

Laboratory 4

End of Lab Questions

Name ________________________________   Lab Section ______________

1. Name the specific structure that produces the spores of the moss. ________________

2. Name the green-leafy portion of the moss. ________________

3. Specifically, what are the filaments that grow out of the moss spore called? ________

4. Describe the differences between the antheridia and the archegonia.

5. What is the difference between the gametophyte and the sporophyte?

6. Describe where liverworts grow naturally.

Laboratory 5

End of Lab Questions

Name ________________________________   Lab Section ____________

1. What are the clusters of sporangia called on the backside of a frond? __________

2. What is an underground stem called? __________________

3. What is the proper name of the fern gametophyte? __________

4. What is so unusual about the whisk fern?

5. Is a club moss really a moss? Why or why not.

6. After touching the horsetail, why do you think the Native Americans call it scouring rush?

Laboratory 6

End of Lab Questions

Name _________________________________   Lab Section _______________

1. What is a mature ovule with a fertilized egg called? _______________________

2. How many nuclei are in a mature ovule? _______________________________

3. In the seed, the entire new plant grows from what? ______________________

4. Name three organisms, other than insects, that pollinate flowers (from the movie).

5. How many pollen grains are produced from a single microsporocyte?
    a. 1
    b. 2
    c. 3
    d. 4

6. What are all the petals called, collectively, and what are all the sepals called, collectively?

7. Matching

    a. ______ Pollen tetrad   1. Female portion of the flower

    b. ______ Pistil   2. Male portion of the flower

    c. ______ Stigma   3. None of the above

    d. ______ Microsporocyte

    e ______ Anther

    f. ______ Ovary

    g. ______ Ovule

    h. ______ Egg

    i. ______ Calyx

Laboratory 7

End of Lab Questions

Name _________________________________ Lab Section ____________

1. What is the primary function of the leaves? ________________________________

2. What structure's function is to increase the surface area of the root? ____________

3. What will the bud between the developing leaves become? __________________

4. Why do roots have a root cap and stems do not?

5. Which of the following is not a function of the root?

    1. anchorage
    2. photosynthesis
    3. storage
    4. absorption

6. Sketch and show the relative position of vascular tissue in each of the following, a root, a dicot stem and a monocot stem.

7. Matching – Where are the following found?

    a. ______ Palisade layer   1. Root

    b. ______ Cortex   2. Stem

    c. ______ Epidermis   3. Leaves

    d. ______ Root hairs

    e ______ Stomata

    f. ______ Region of maturation

    g. ______ Vascular tissue

    h. ______ Cuticle

Laboratory 8

End of Lab Questions

Name ____ Lab Section _____________

1. What structures are for support and protection in the sponge? _____________ and

   ____

2. What does the word sessile mean? ____

3. Name the opening that water enters the sponge through. _________________________

4. What are the three functions of a collar cell?

5. Explain why it is important for water to continue to be circulated through the sponge. (Two reasons)

6. What are the two types of asexual reproduction that a sponge can do?

Laboratory 9

End of Lab Questions

Name ________________________________   Lab Section ____________

1. The Petoskey stone is really fossilized _________________.

2. What is the planula? _____________________________________

3. What are nematocysts? _______________________

4. What are the two body types of the Cnidarians and how are they different from one another?

5. Explain how the *Hydra* reproduces asexually.

6. Name the two different types of polyps in the *Obelia* and how can you tell them apart?

Laboratory 10

End of Lab Questions

Name _________________________________   Lab Section _____________

1. Are rotifers multicellular or single celled? _________________

2. What do the rotifers eat? _________________________________

3. What are the structures to one end that look like they are rotating? _____________

4. Where can one find rotifers in nature?

5. What are some of the advantages of being multicellular verses being single celled? (use
    your textbook)

Laboratory 11

End of Lab Questions

Name ________________________________   Lab Section _____________

1. How is a clam different from other mollusks?

2. How do you tell the age of a clam?

3. Explain how a clam gets its nutrients.

4. Do clams have an incomplete or complete digestive system?

Laboratory 12

End of Lab Questions

Name ___________________________________   Lab Section _____________

1. How do you get Trichinosis? ____

2. Where do planaria live? ____

3. Which worm that we looked at has no digestive system? _______________

4. Why do we not have all of these worm parasites in Michigan?

5. Matching

    a. _____Pinworm   1. Platyhelminthes

    b. _____Earthworm   2. Nematoda

    c. _____Segmented worm   3. Annelida

    d. _____Swimmer's itch

    e. _____Vinegar eel

    f. _____Planaria

6. All flukes are

    a. Roundworms

    b. Parasites

    c. Beneficial

    d. Not contagious

7. Define the following structures:

    a. crop

    b. intestine

    c. nephridium

    d. ventral nerve cord

Laboratory 13

End of Lab Questions

Name ___________________________________   Lab Section ______________

1. The grasshoppers that we looked at are _____________________ grasshoppers.

2. How many pairs of appendages does the crayfish have? _____________________

3. Name the air holes that are found on the abdomen and thorax of the grasshopper.

   ____

4. Name the body regions of both, insects and crayfish.

5. Explain how you can tell the female grasshopper from the male grasshopper.

6. Which of the following structures is not found on the crayfish?

   a. green gland

   b. intestine

   c. heart

   d. gastric caeca

Laboratory 14

End of Lab Questions

Name _________________________________   Lab Section ____________

1. What structures are common in all chordates? _________________ and _________________

2. What is the most obvious region in the frog's brain? _________________________

3. If your frog has gray and white material in its abdomen, is your frog male or female? _________________ What is this material? _________________________

4. Explain how the invertebrate and vertebrates differ?

5. What class does the frog belong to

    a. Fish

    b. Amphibian

    c. Reptile

    d. Bird

6. Define the following structures:

    a. Tympanum

    b. Cloaca

    c. Cirri

    d. Spleen

Laboratory 15

End of Lab Questions

Name ____   Lab Section ______________

1. Explain how mammals are different from the other major vertebrate chordates.

2. What is the function of the following organs that are found in both humans and pigs?

    a. Heart

    b. Stomach

    c. Colon

    d. Trachea

    e. Ovary

    f. Adrenal gland

3. Make a list of the major organs of the digestive system that food would pass through starting with the mouth, ending with the anus in a pig. (You don't have to actually draw it)

Laboratory 16

End of Lab Questions

Name ___________________________________   Lab Section ______________

1. How are the three distinct groups of fish different from one another?

2. Explain the dental formula for an average adult human.

3. Turtles use their _________________ for some of the support and protection that the

endoskeleton normal does.

4. Which of the following organisms do not have true teeth but denticles?

       a. cat

       b. shark

       c. woodchuck

       d. snake

       e. bat

5. Describe the differences between; insectivore, carnivore, herbivore and rodent.